Tru

Traveling Bags, and Satchels

Price Guide by
Roseann Ettinger

4880 Lower Valley Road, Atglen, PA 19310 USA

Library of Congress Catalog Card Number: 98-86372

Design by Blair Loughrey
Type set in Galleria/Aldine 721 Lt Bt

ISBN: 0-7643-0617-0
Printed in China

Published by Schiffer Publishing Ltd.
4880 Lower Valley Road
Atglen, PA 19310
Phone: (610) 593-1777; Fax: (610) 593-2002
E-mail: Schifferbk@aol.com
Please write for a free catalog.
This book may be purchased from the publisher.
Please include $3.95 for shipping.

In Europe, Schiffer books are distributed by
Bushwood Books
6 Marksbury Avenue
Kew Gardens
Surrey TW9 4JF England
Phone: 44(0)181-392-8585; Fax: 44(0) 181-392-9876
E-mail: Bushwd@aol.com

Please try your bookstore first.

We are interested in hearing from authors
with book ideas on related subjects.

Contents

Trunks

Metal Covered

No. 50.

SIZES, 26 in., 28 in., 30 in., 32 in., 34 in., 36 in.

Crystal covered, iron bottom, patent veneer top, wrought top corner bumpers.

No. 10.

SIZES, **26 in., 28 in., 30 in., 32 in., 34 in., 36 in.**

Crystal covered, iron bottom, patent veneer tops, valance all around.

No. 53.

SIZES, **28 in.,** **30 in.,** **32 in.,** **34 in.,** **36 in.**

Crystal covered, iron bottom, patent veneer top, valance all around, malleable corner clamps.

No. 100.

SIZES, **28 in.,** **30 in.,** **32 in.,** **34 in.,** **36 in.**

Crystal covered, iron bottom, patent veneer top, valance all around, wrought clamps at all outside corners.

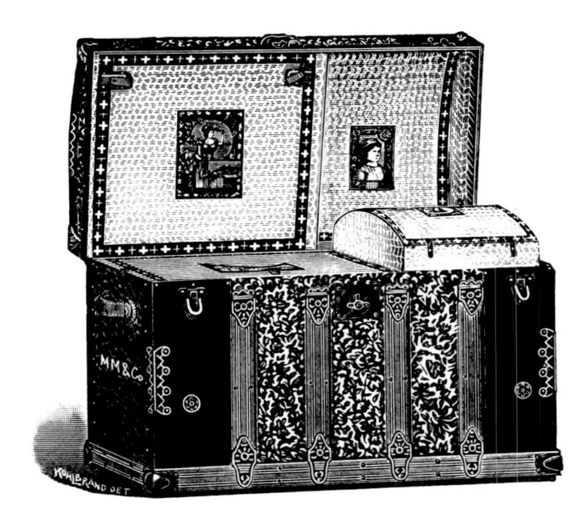

No. 125.

SIZES, 28 in., 30 in., 32 in., 34 in., 36 in.

Crystal covered, iron bottom, patent veneer tops, valance all around, malleable corner clamps, wrought outside slat clamps.

No. 200.

SIZES, 28 in., 30 in., 32 in., 34 in., 36 in.

Crystal covered, iron bottom, patent veneer tops, valance all around; all slat guards and corner clamps are malleable iron; colored facings on tray and top.

No. 42.

SIZES, 28 in., 30 in., 32 in., 34 in., 36 in.

Crystal covered, iron bottom, patent veneer tops, valance all around, malleable iron slat tip clamps and corner bumpers, wrought outside slat clamps.

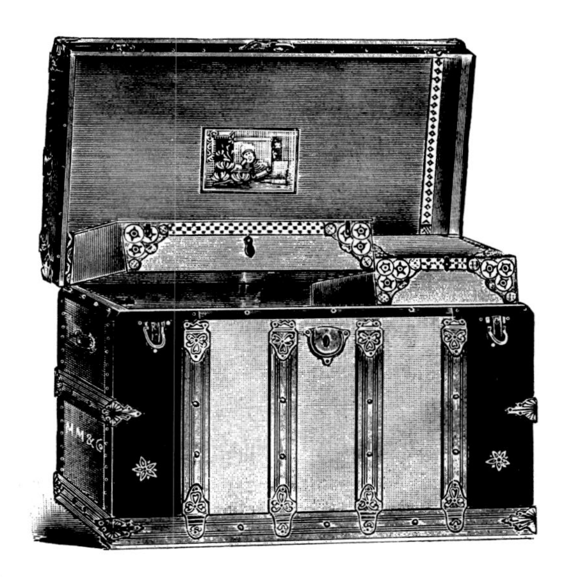

No. 210.

SIZES, 28 in., 30 in., 32 in., 34 in.

Imitation canvas covered, heavy malleable corner and outside slat clamps, sheet iron bottom; tasty interior, faced in colors.

No. 156.

SIZES, **30 in.,** **32 in.,** **34 in.,** **36 in.**

Crystal covered, iron bottom, patent veneer tops, valance all around, malleable and wrought iron corner and outside slat clamps. A great big trunk, cheap and good.

No. 47.

SIZES, 28 in., 30 in., 32 in., 34 in., 36 in., 38 in.

Crystal covered, iron bottom, patent veneer tops. Malleable iron outside clamps, slat tip clamps and top edge binding. Brass lock. M. Maier's patent corner hinge, bumper and stay combined.

No. 61.

SIZES, 32 in., 34 in., 36 in.

Crystal covered. Iron bottom. Patent veneer tops. Malleable slat tip, corner and slat clamps. M. Maier's patent corner hinge, bumper and stay combined. Swing tray.

No. 66.

SIZES, 32 in., 34 in., 36 in.

Crystal covered, patent veneered tops, iron bottom; our own design rich white malleable iron slat guards, corner bumpers, slat and valance clamps; M. M.'s patent corner hinge and stay combined; inside finish cloth faced.

G. P. PACKING.

CRYSTAL PACKING.

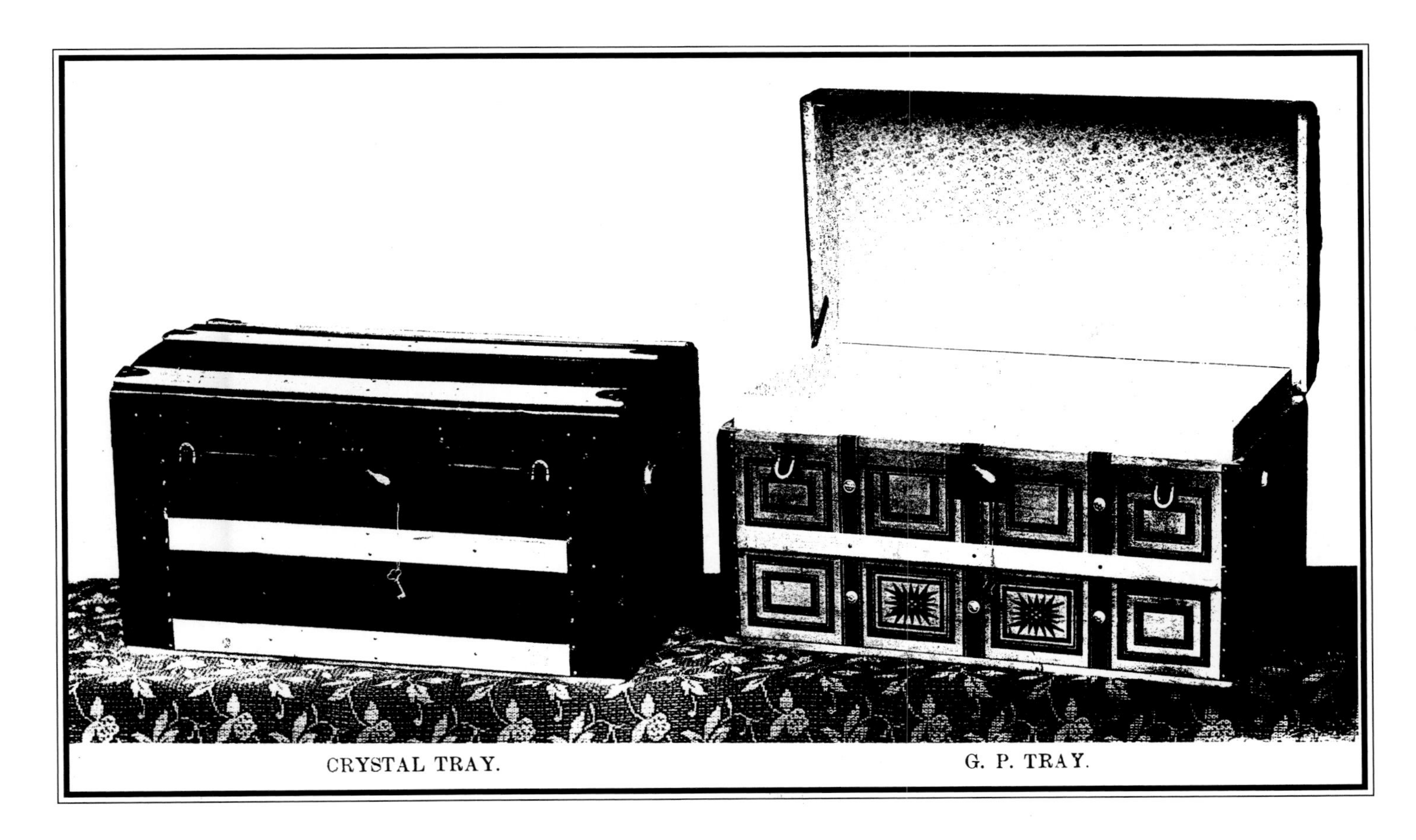

CRYSTAL TRAY.

G. P. TRAY.

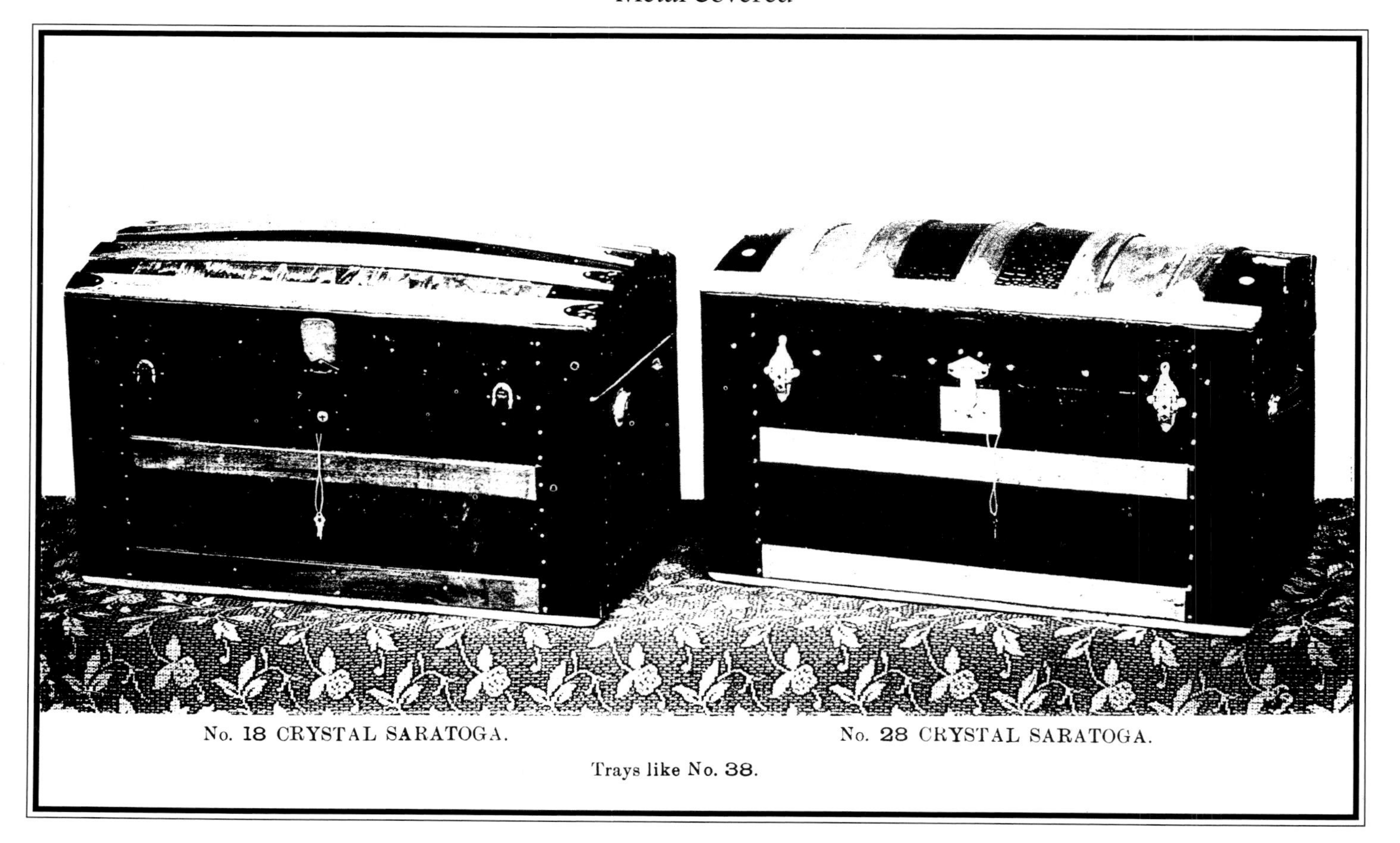

No. 18 CRYSTAL SARATOGA.

No. 28 CRYSTAL SARATOGA.

Trays like No. 38.

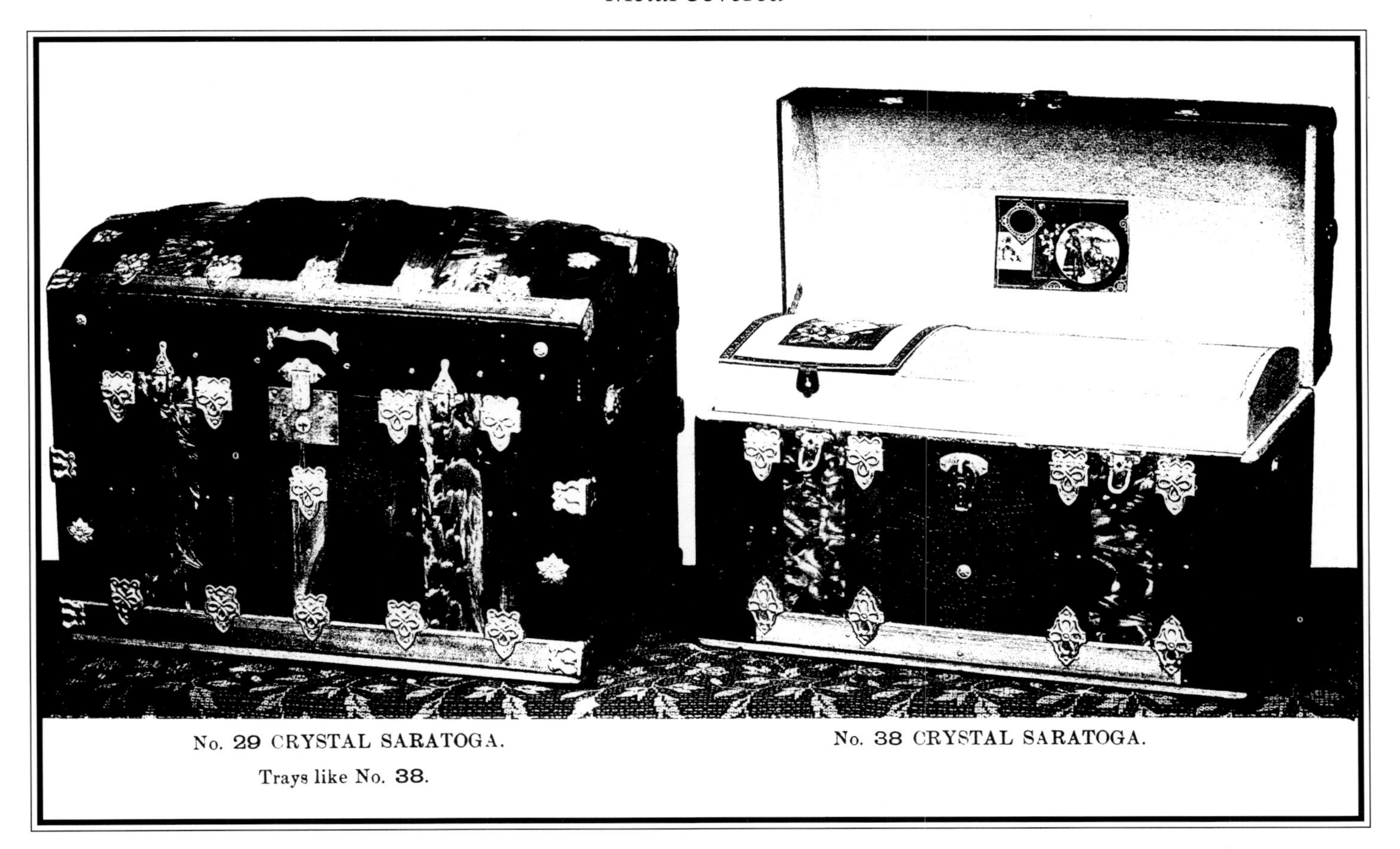

No. 29 CRYSTAL SARATOGA.

Trays like No. 38.

No. 38 CRYSTAL SARATOGA.

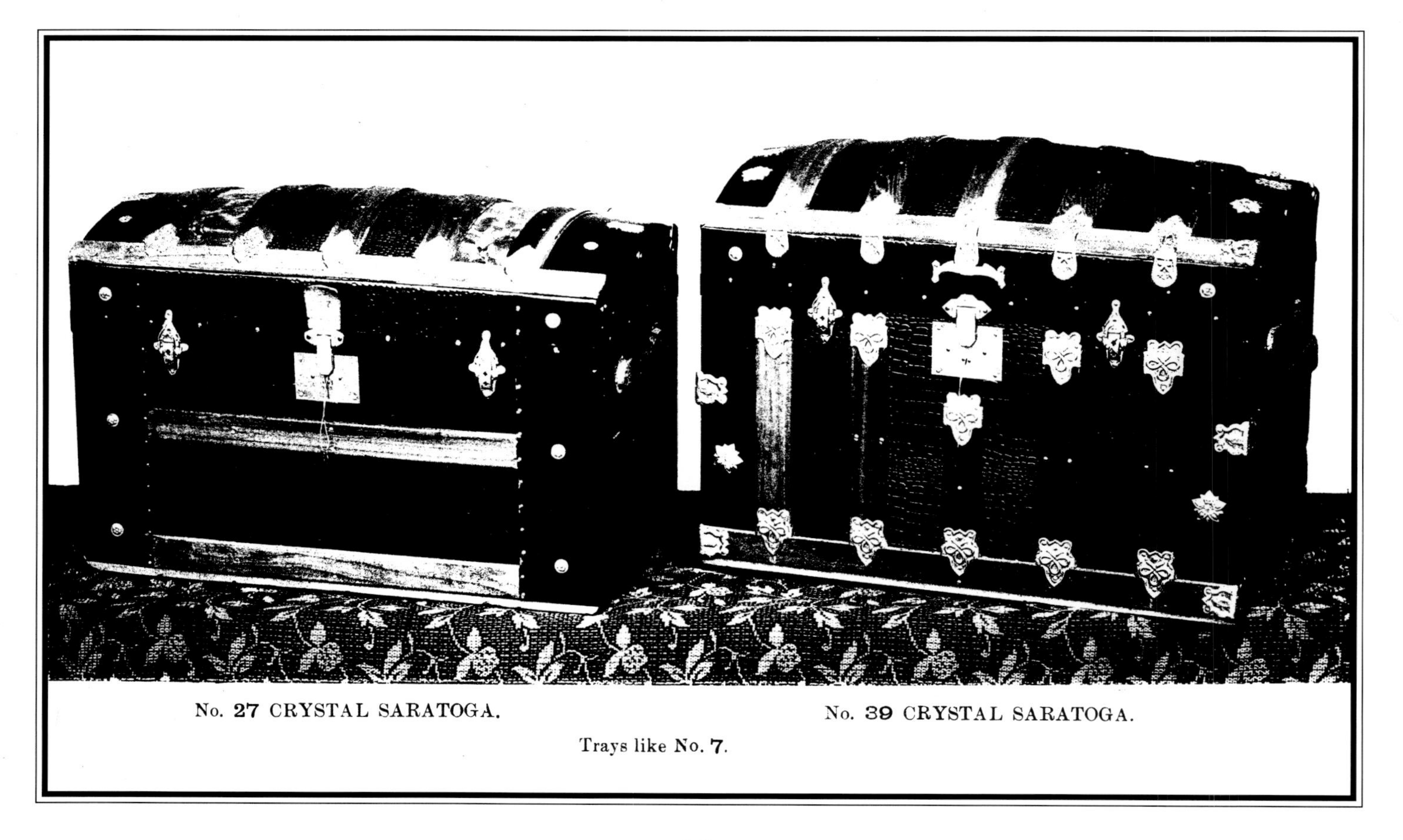

No. 27 CRYSTAL SARATOGA.

No. 39 CRYSTAL SARATOGA.

Trays like No. 7.

$4.55 BEST BARREL TOP TRUNK, STEEL COVERED. PRICES REDUCED

No. 33K1050 Heaviest, Fancy Metal Covered Trunk, with large barrel top, five heavy hardwood barrel stave slats over top and down side, two hardwood slats across each end. Heavy, malleable trimmings and reinforcements throughout. Brass Monitor lock, strongest malleable iron, patent bolts, rollers, hinges, catches, etc. Leather handles, full finished hinged tray with separately covered compartment and bonnet box, also fall-in cover top compartment and extra skirt tray which fits in underneath first tray. Iron bottom, and made and trimmed and reinforced at every danger point exactly as illustrated. Extra large size, thick basswood box. One of the greatest trunks in this style ever offered. We will, at any time, replace any trunk of this number which fails to give satisfaction. Where else can you get a guarantee like this on trunks? **Remember to state the size and catalogue number.**

Five Slats on Cover

Iron Bottom.

Sheet Iron bound

Length, 28 in.	Width, 16½ in.	Height, 22 in.	Weight, 48 lbs.	Reduced price, now $4.55
Length, 32 in.	Width, 18½ in.	Height, 24 in.	Weight, 51 lbs.	Reduced price, now 5.25
Length, 36 in.	Width, 20½ in.	Height, 26 in.	Weight, 62 lbs.	Reduced price, now 5.95

See No. 33K1050 for better barrel top trunk.

Extra Suit or Skirt Tray.

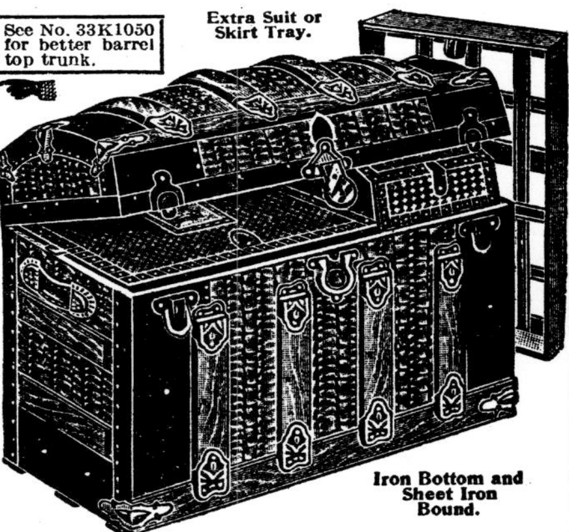

Iron Bottom and Sheet Iron Bound.

$3.65 FANCY METAL COVERED BARREL TOP TRUNK

No. 33K1014 Fancy Metal Covered Trunk, large thick basswood box, paper lined, large barrel top with four heavy hardwood bar slats over top and on side, and two across each end. Sheet iron bound, and heavy fancy malleable trimmings and reinforcements. Iron bottom, patent bar bolts and rollers, heavy steel hinges and strong end clamps. Strong brass Monitor lock, fancy catches and leather handles, and contains tray with bonnet box and side compartment separately covered, also fall-in covered top compartment. Has skirt tray, which fits in underneath the upper tray and when not in use can be inverted and takes up comparatively no room. This is the greatest value ever offered in a low priced barrel top trunk, and we guarantee it to give satisfaction. **Be sure to state the size wanted.**

Length, 28 in.	Width, 16½ in.	Height, 20 in.	Weight, 41 lbs.	Price..........$3.65
Length, 32 in.	Width, 18½ in.	Height, 22 in.	Weight, 49 lbs.	Price.......... 4.25
Length, 36 in.	Width, 20½ in.	Height, 24 in.	Weight, 60 lbs.	Price.......... 4.85

$1.85 CRYSTALLIZED METAL COVERED TRUNK, BARREL TOP. Reduced Prices.

Strong and Serviceable. Full Size.

Iron Bottom.

No. 33K1002 Substantially made Barrel Top Trunk, with four hardwood slats over top and two slats on sides and one on each end. Sheet iron bound, japanned steel end clamps, iron bottom, special bar bolts, hinges, rollers and catches and strong hasp lock and leather handles. Contains set up tray with side compartment and covered bonnet box. A full size trunk at a very low price. Cannot be duplicated at $2.00 more than what we ask. But we honestly advise the purchase of a better trunk, as a good trunk lasts many years and always insures safe transportation to its contents, and we recommend to you our No. 33K1014 or No. 33K1050, illustrated and described at bottom of this page. Remember though, for the price, this is the best trunk ever made. **State size wanted and give correct catalogue number.**

All prices reduced.

Length, 26 in.	Width, 14½ in.	Height, 17½ in.	Weight, 28 lbs.	Reduced price, $1.85
Length, 30 in.	Width, 16½ in.	Height, 19½ in.	Weight, 35 lbs.	Reduced price, 2.45
Length, 34 in.	Width, 18½ in.	Height, 21½ in.	Weight, 42 lbs.	Reduced price, 3.05
Length, 36 in.	Width, 19½ in.	Height, 22½ in.	Weight, 47 lbs.	Reduced price, 3.35

$4.95 HEAVIEST MONITOR TOP BLACK ENAMEL STEEL COVERED TRUNK.

ONE OF THE HANDSOMEST and STRONGEST TRUNKS MADE

No. 33K1040 Black Enamel Steel Covered Monitor Top Trunk. Four heavy hardwood slats over top and down side, two across each end and one lengthwise across top, intersecting cross slats. Extra heavy brass malleable clamps and reinforcements, brass Excelsior lock, heavy patent bolts, rollers, hinges, catches, stitched leather handles, etc. Contains tray with bonnet box and side compartment, fall-in-top compartment in cover, all separately covered. Has extra skirt tray which can be inverted and takes up comparatively no room. Note the heavy malleable trimmings, the number of and width and thickness of the hardwood slats, the strong reinforcements, and remember that this trunk has an iron bottom and is made on extra large size thick basswood box, one of the handsomest and most serviceable trunks made. We are so confident of the strength and the superiority of this trunk that we will, at any time, replace any one that does not give absolute satisfaction. Where is the dealer that offers this guarantee on any trunk he sells? **Be sure to state size and catalogue number.**

Length, 28 in.	Width, 17 in.	Height, 20 in.	Weight, 45 lbs.	Special price....$4.95
Length, 32 in.	Width, 19 in.	Height, 22 in.	Weight, 55 lbs.	Special price.... 5.65
Length, 36 in.	Width, 21 in.	Height, 24 in.	Weight, 64 lbs.	Special price.... 6.35
Length, 38 in.	Width, 22 in.	Height, 25 in.	Weight, 70 lbs.	Special price.... 6.70

$3.65 MONITOR TOP STEEL COVERED TRUNK. Prices Reduced

Skirt Tray.

Iron Bottom.

No. 33K1022 Handsome Steel Covered Black Enamel Monitor Top Trunk. Large box made of thick basswood, paper lined, has flat top with rounded edges and is finished with four heavy hardwood bar slats over top and two on sides and around ends. Heavy sheet iron bottom and is thoroughly reinforced and trimmed with metal trimmings, steel end clamps, patent bar bolts, heavy steel hinges and rollers, brass monitor lock and catches, leather handles, and contains tray with separately covered bonnet box and side compartment, also skirt tray which fits underneath first tray and when desired can be inverted and takes up comparatively no room. A trunk that is easily worth $2.00 to $3.00 more than our low price. We gladly offer this trunk as proof of our assertion of absolutely the best trunks ever made at prices which cannot be met. Guaranteed to give absolute satisfaction in every respect or your money back. **Be sure to state size wanted.** All prices REDUCED as follows:

Length, 28 in.	Width, 16 in.	Height, 18½ in.	Weight, 37 lbs.	Reduced price.. $3.65
Length, 32 in.	Width, 18 in.	Height, 20½ in.	Weight, 47 lbs.	Reduced price.. 4.25
Length 36 in.	Width, 20 in.	Height, 22½ in.	Weight, 58 lbs.	Reduced price.. 4.85

WOOD & FIBER

No. 34.

SIZES, **28 in., 30 in., 32 in., 34 in., 36 in., 38 in.**

Rustic; covered with hardwood slats; white malleable iron trimmings of our own design; M. M.'s patent corner hinges and stay combined; slats all tipped with iron; hinged tray; lining in body and tray facing of cloth.

BASKET TRUNK.

SIZES, **30 in.,** **33 in.,** **36 in.,** **39 in.**

Made from select willows reinforced with rattan. These baskets are made especially for us and of special proportions. Covering of best quality japanned duck ; corner binding of heavy grain leather, hand stitched ; fine brass Excelsior lock; hickory cleat on back, hickory cleats on bottom, with steel shoes at ends; inside lined with Irish linen, top lined with imitation leather, padded. Straps. handles, cleats and trimmings secured to basket with heavy rivets, rivetted to special made burrs. We claim this to be the best made basket trunk in the market.

No. 5 G. P. SARATOGA.

No. 25 G. P. SARATOGA.

No. 6 G. P. SARATOGA.

Trays like No. 5 G. P.

No. 4 G. P. SARATOGA.

No. 33 G. P. SARATOGA.

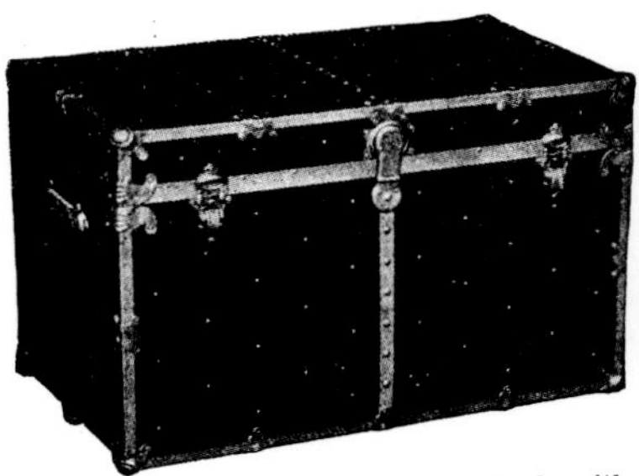

Green sheet steel covered, studded with nails; bronzed steel binding and center band brass plated locks, bolts, corners and clamps; fancy print lining; top and dress tray; covered top tray divided into two compartments.

		34-inch	36-inch	38-inch	40-inch
No. 3016	Each	**$23.30**	**$29.00**	**$28.70**	**$30.40**

Same as above, steamer style. **Black.**

		34-inch	36-inch	38-inch
No. 3506	Each	**$19.50**	**$21.25**	**$21.90**

FIBRE COVERED TRUNK

Black vulcanized fibre covering. Bronzed steel binding. Two bronzed steel center bands. Brass plated lock, bolts, dowels, corners and other trimmings. Fancy print lining. Covered tray divided into two compartments. Extra dress tray.

		34-inch	36-inch	38-inch
No. 4606	Each	**$29.65**	**$30.40**	**$31.35**

Same as above, steamer style.

		34-inch	36-inch	38-inch
No. 4406	Each	**$19.45**	**$21.20**	**$21.90**

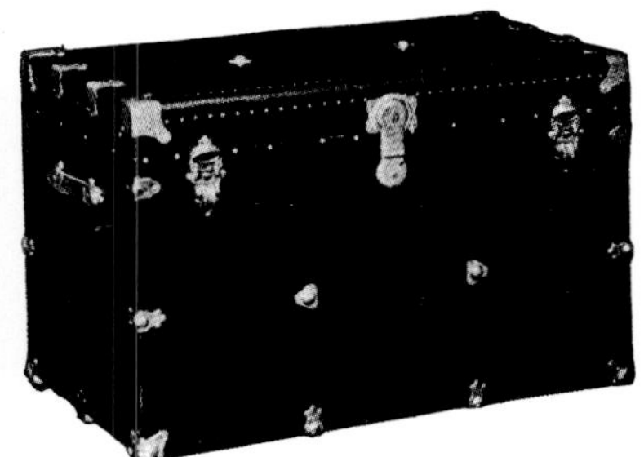

Black fibre covered; basswood box; rounded edges; maroon fibre binding; brass plated lock, bolts, corners and clamps; fancy print lining; two trays; covered top tray divided into two compartments.

		34-inch	36-inch	38-inch
No. 8046	Each	**$29.90**	**$30.80**	**$31.60**

Same as above, steamer style; maroon fibre covering.

		34-inch	36-inch	38-inch
No. 8606	Each	**$21.95**	**$22.60**	**$23.30**

Dark green fibre covered trunk. Round edge. Green fibre binding. Brass plated steel Excelsior lock, bolts, dowels, corners, clamps and other trimmings. Cloth lined throughout. Covered top tray divided into two compartments. Extra dress tray.

		34-inch	36-inch	38-inch	40-inch
No. 8036	Each	**$37.90**	**$39.25**	**$40.75**	**$42.00**

Same as above, steamer style.

		34-inch	36-inch	38-inch	40-inch
No. 8706	Each	**$27.75**	**$28.70**	**$29.70**	**$30.65**

FIBRE COVERED SLATLESS TRUNK

Box made of 3-ply basswood. Vulcanized fibre covered. Extra heavy fibre binding; brass plated steel trimmings; brass plated Excelsior lock; extra large, heavy corner bumpers; top tray, with folding lid; extra dress tray; fancy print lining.

		34-inch	36-inch	38-inch	40-inch
No. 5006	Each	**$34.65**	**$35.05**	**$37.45**	**$38.85**

Same as above, steamer style.

		34-inch	36-inch	38-inch	40-inch
No. 4486	Each	**$31.50**	**$32.85**	**$34.25**	**$35.65**

Round Edge Vulcanized Fibre Covered Trunk

Dark green color. Vulcanized fibre binding. All edges rounded. Brass plated lock, bolts, dowels, corners and clamps. Cretonne lined. Covered top tray divided into two compartments. Extra dress tray.

		36-inch	38-inch	40-inch
No. 4256	Each	**$49.95**	**$50.90**	**$52.70**
Same as above, steamer style.				
No. 4206	Each	**40.20**	**41.60**	**43.00**

Canvas Covered

No. 54.

SIZES, **28 in.,** **30 in.,** **32 in.,** **34 in.,** **36 in.**

Canvas covered, well painted. Iron bottom. Wrought steel clamps at every slat.

No. 57.

SIZES, 28 in., 30 in., 32 in., 34 in., 36 in.

Canvas covered, well painted. Iron bottom. Corner rollers, wrought steel clamps at every slat.

No. 58.

SIZES, **28 in., 30 in., 32 in., 34 in., 36 in.**

Duck covered, well painted, iron bottom, corner rollers, steel corner bumpers. Wrought steel clamps at every slat.

No. 59.

SIZES, **28 in., 30 in., 32 in., 34 in., 36 in.**

Duck covered, well painted, iron bottom, malleable corner rollers, steel corner bumpers and slat clamps, sliding handles, brass Excelsior lock. Cloth faced tray and top. Dress tray.

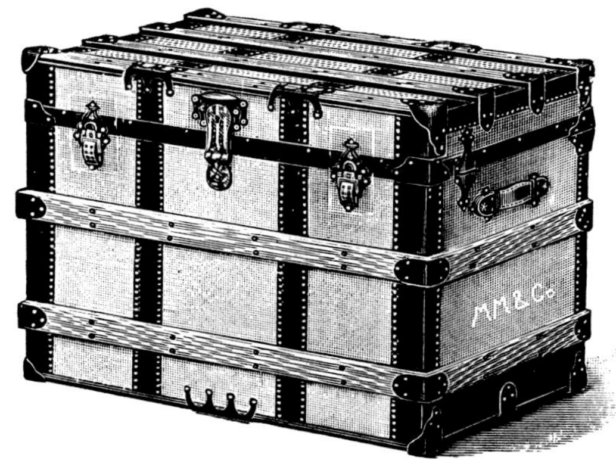

No. 6.

SIZES, 26 in., 28 in., 30 in., 32 in., 34 in., 36 in.

Duck covered, well painted. Iron bottom. Malleable corner bumpers of our own design. M. Maier's patent corner hinge and stay combined, heavy strap hinge in centre. Wrought steel clamps and valance guards. Heavy dowel bolt on end. Malleable chain clamps top and bottom. Tray and top full cloth lined. Suitable trunk for gent or lady, as professional or sample trunk.

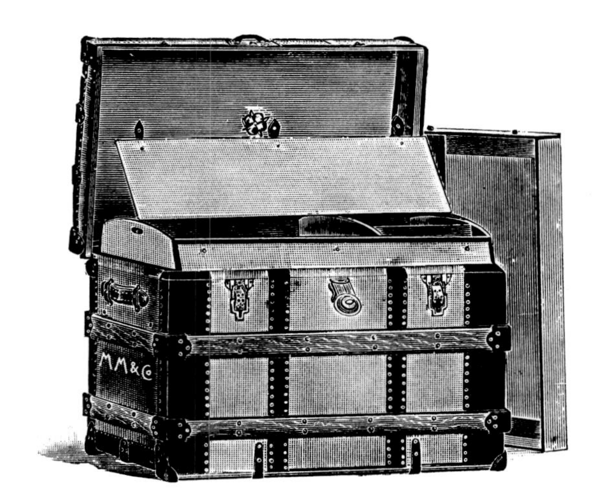

No. 120.

SIZES, 26 in., 28 in., 30 in., 32 in., 34 in., 36 in.

Duck covered, well painted, iron bottom, malleable corner bumpers of our own design top and bottom. Wrought steel slat clamps and knees. M. M.'s patent corner hinge and stay combined. Slats all tipped and bound. Trunk cloth lined throughout. Dress tray.

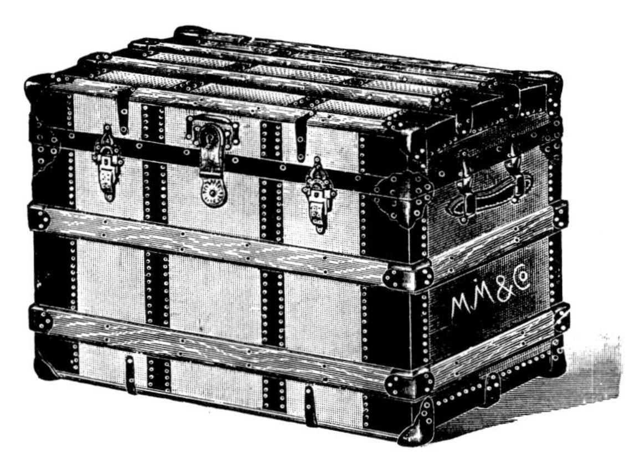

No. 85.

SIZES, **28 in., 30 in., 32 in., 34 in., 36 in., 38 in.**

Duck covered, well painted. Malleable corner bumpers of our own design top and bottom. Doweled handle caps, heavy strap hinges, steel mouth clamps and valance clamps. Wrought steel slat clamps and knees. Slats all tipped and bound. Hinges, bolts, dowels, etc., riveted with heavy burrs inside. Trunk cloth lined throughout. Dress tray.

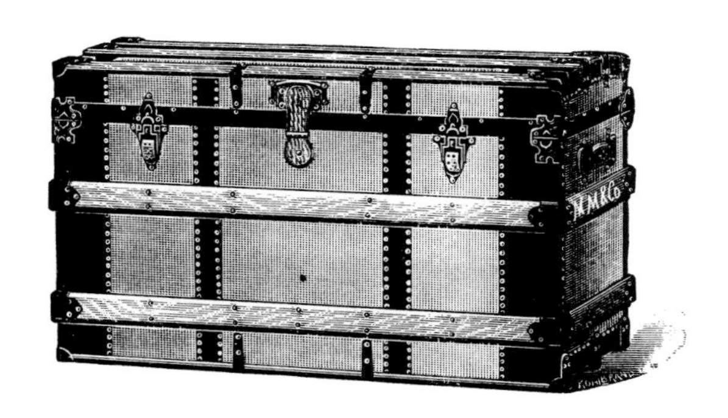

No. 35.

SIZES, 36 in., 38 in., 40 in., 42 in., 44 in.

Duck covered, well painted, all hardwood slats, M. Maier's patent corner hinge, guard and stay combined. Patent doweled mouth and valance clamps. Movable partitions in bottom of body, making apartments for bonnets, shoes or linen. Trunk lined throughout with cloth.

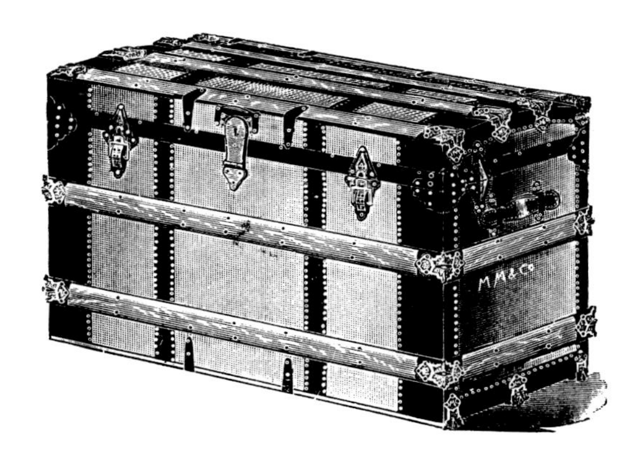

No. 90.

SIZES, 36 in., 38 in., 40 in., 42 in., 44 in.

Duck covered dress trunk. Well painted. Malleable slat and corner clamps of our own design. Steel mouth and valance clamps all around. Heavy strap hinges. Movable cross partitions in tray. Trunk lined throughout with cloth. Three dress trays.

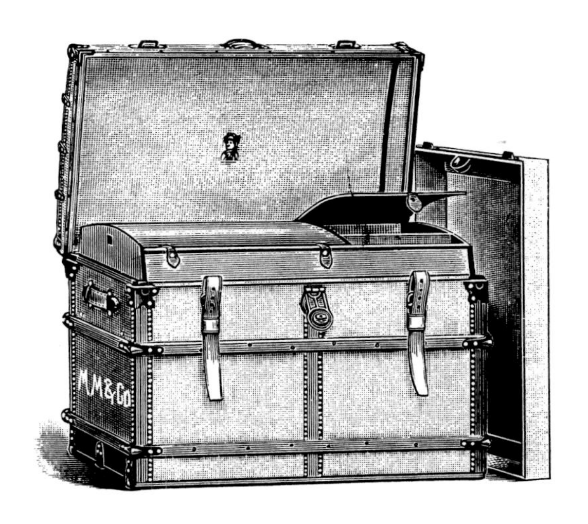

No. 70.

SIZES, 28 in., 30 in., 32 in., 34 in., 36 in.

Duck covered, well painted. Malleable lip clamps of our own design at every slat. M. M.'s patent corner hinge guard and stay and doweled mouth and valance clamp on front. Leather corner binding and centre band. Cloth lined throughout. Dress tray.

No. 75.

SIZES, **28 in., 30 in., 32 in., 34 in., 36 in.**

Duck covered, well painted. Best quality oak tanned leather binding and centre band. Polished brassed trimmings. Cloth lined throughout. Dress tray.

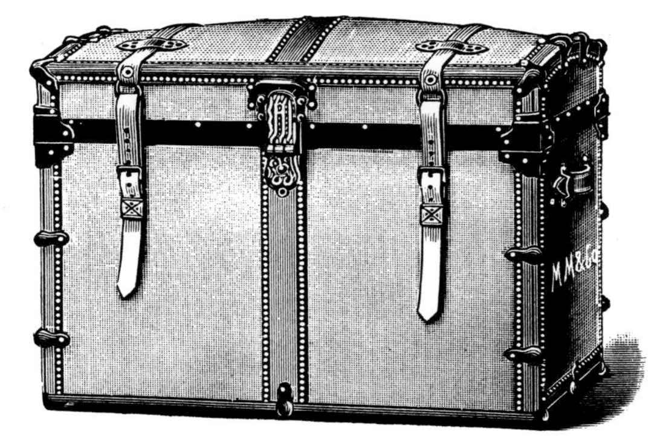

No. 18.

SIZES, 30 in., 32 in., 34 in., 36 in.,

Duck covered, well painted. Best oak tanned leather binding centre bands and straps. Trimmings malleable and of our own design. M. M.'s patent corner hinge. Box shoulder and valance clamp on front. Cloth lined throughout. Tray lids leather bound. Dress tray.

No. 80.

SIZES, 30 in., 32 in., 34 in., 36 in.

Duck covered, well painted, best oak tanned leather binding, centre band and straps, solid brassed malleable trimmings, three heavy strap hinges on back, cloth lined throughout, dress tray.

No. 20.

SIZES, **30 in.,** **32 in.,** **34 in.,** **36 in.**

Duck covered, well painted, best oak tanned leather binding, centre band and straps. Trimmings are solid cast brass of our own design. Box shoulder and valance clamps back and front. Many of the trimmings secured by solid rivets and burrs. Cloth lined throughout, padded silk hat rest in tray, tray lids leather bound, dress tray.

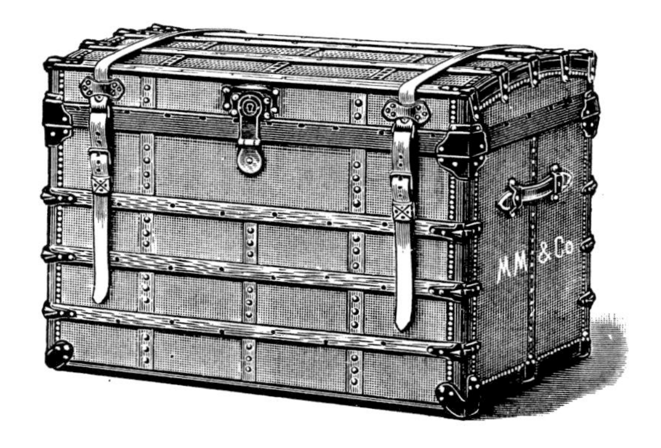

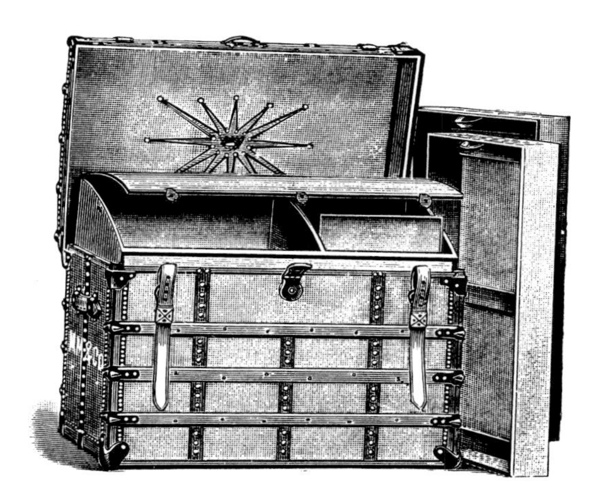

No. 144.

SIZES, 32 in., 34 in., 36 in., 38 in.

Duck covered, well painted, large box, all hickory slats, heavy leather corner binding. All bands and valance duck covered. Trimmings are malleable, our own design clamps with lips, M. M's patent corner guard, hinge and stay. Box shoulder and valance clamp on front. All clamps and trimmings secured with solid rivets and burrs, Irish linen lined throughout, movable partitions in tray, two dress trays with straps, very light trunk, basswood box.

No. 1.

SIZES, 30 in., 32 in., 34 in., 36 in.

Duck covered, well painted, extra made basswood box, best russet oak tanned leather in binding centre bands and straps, solid cast brass trimmings of our own design, box shoulder and valance clamps back and front, trimmings secured with solid rivets and burrs, hickory bottom cleats with ½ inch heavy iron protector on outside edge, heavy double back strap hinges, Irish linen lined throughout, tray lids leather bound, removable padded silk hat rest in hat box, dress tray with web ties.

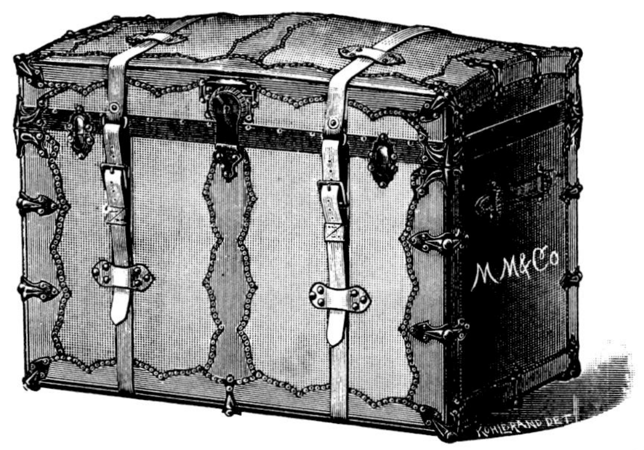

No. 400.

SIZES, 30 in., 32 in., 34 in., 36 in.

Duck covered, well painted, extra made basswood box, best oak tanned leather binding, centre band and straps; solid bronze metal trimmings of our own design, box shoulder and valance clamps back and front, duck covered valance, trimmings all secured with solid rivets and burrs, heavy double backed strap hinges, hickory bottom cleats with heavy ½ inch iron protector on outside edge. Lined with Irish linen throughout, tray lids leather bound.

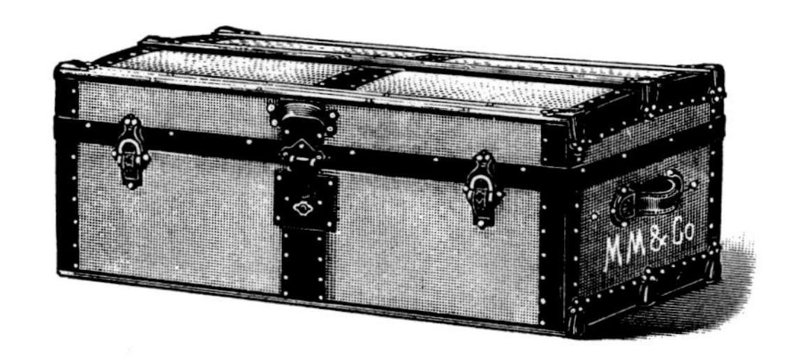

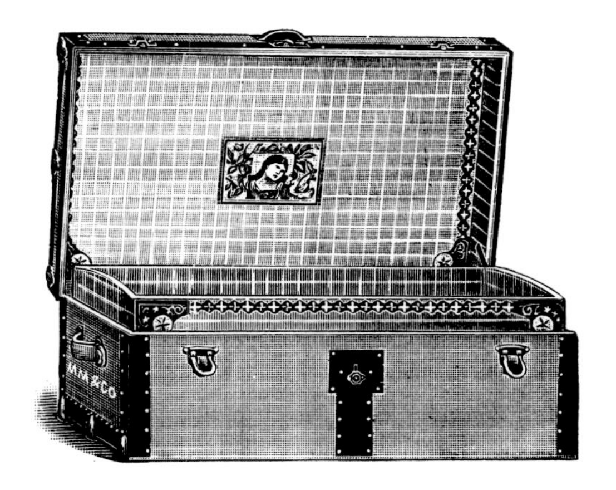

No. 3 STEAMER.

SIZES, 30 in., 32 in., 34 in.

Duck covered, well painted, wrought clamps at every slat, sheet iron bottom.

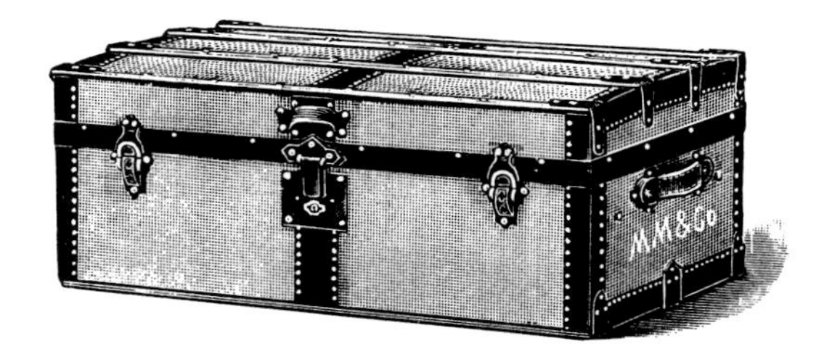

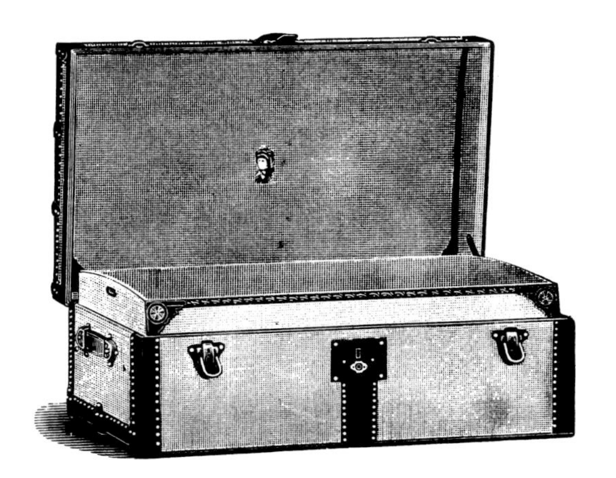

No. 2 STEAMER.

SIZES, **30 in., 32 in., 34 in., 36 in.**

Duck covered, well painted. Wrought steel clamps at every slat. Cloth lined throughout. Iron bottom.

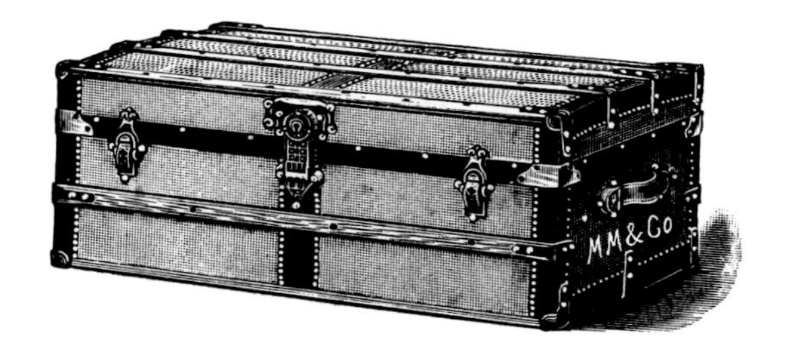

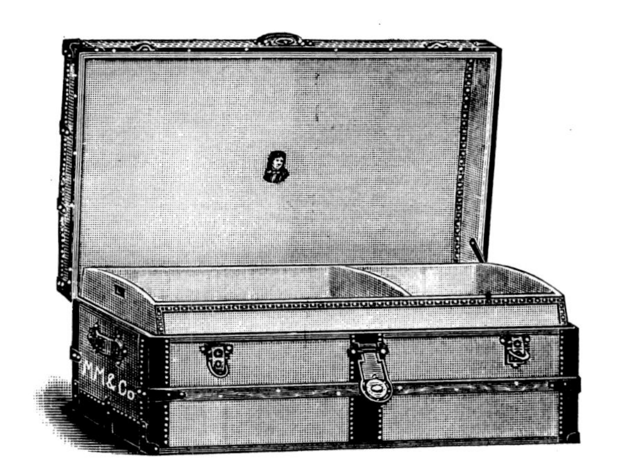

No. 5 STEAMER.

SIZES, **30 in.**, **32 in.**, **34 in.**, **36 in.**

Duck covered, well painted. Brassed corner and slat clamps. Cloth lined throughout.

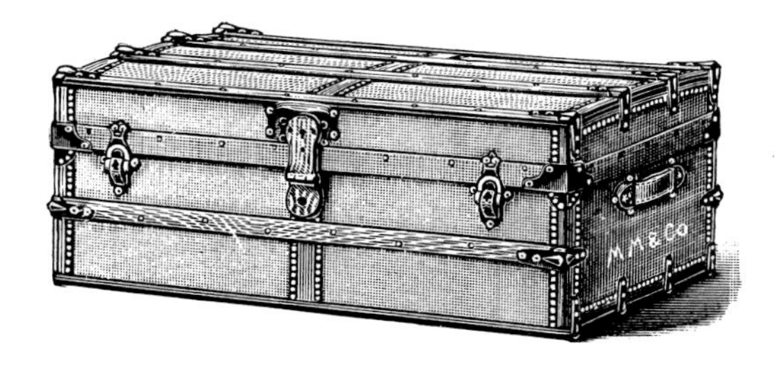

No. 11 STEAMER.

SIZES, **30 in.,** **32 in.,** **34 in.,** **36 in.**

Duck covered, well painted. Duck covered bottom and valance. Neat malleable lip clamps of our own design at every slat. Heavy russet leather corner binding and centre band. Cloth lined throughout.

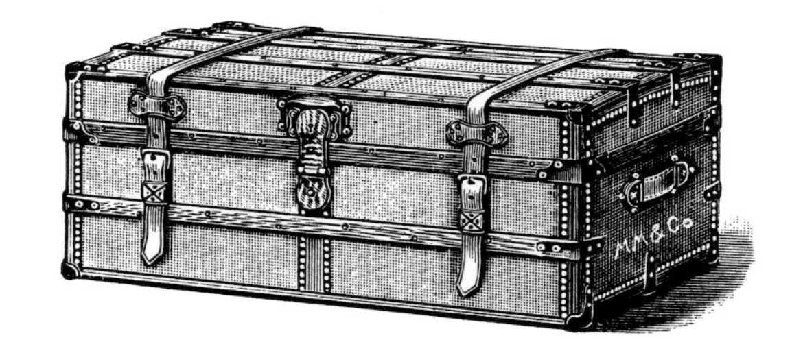

No. 13 STEAMER.

SIZES, **30 in.,** **32 in.,** **34 in.,** **36 in.**

Duck covered, well painted. Duck covered bottom. Heavy russet leather corner binding and centre band. Polished brassed slat, corner and valance clamps. Linen bottom tray. Trunk lined with Irish linen throughout.

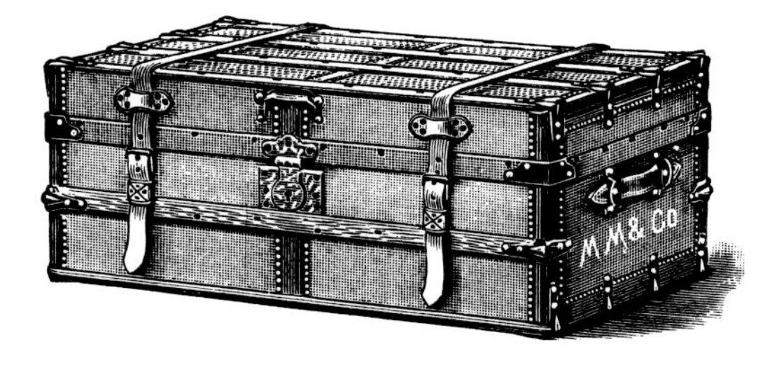

No. 15 STEAMER.

SIZES, 30 in., 32 in., 34 in., 36 in.

Duck covered, well painted. Duck covered bottom and valance. Heavy russet corner binding and centre band. Finely polished cast bronze metal trimmings of our own design at every slat. Heavy strap hinges. Inside Irish linen lined throughout. A handsome and durable trunk.

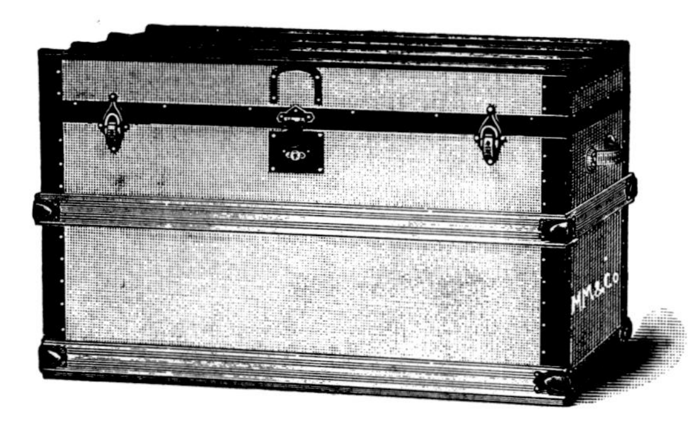

DUCK COVERED PACKER.

PACKING TRUNKS.

Paper packing trunks, covered with imitation leather.

SIZES, **24 in., 28 in., 32 in., 36 in., 40 in.**

Separate sizes, or in full nests.

ROUND TOP.

Imitation leather covered, deep set up tray, hat box, rollers, hasp lock, slats full length of trunk.

SIZES, **26 in., 28 in., 30 in., 32 in., 34 in.**

DUCK COVERED PACKERS.

A large, strong packing trunk, superior to paper packer in size, durability and appearance.

SIZES, **28 in., 30 in., 32 in., 34 in., 36 in., 38 in., 40 in.**

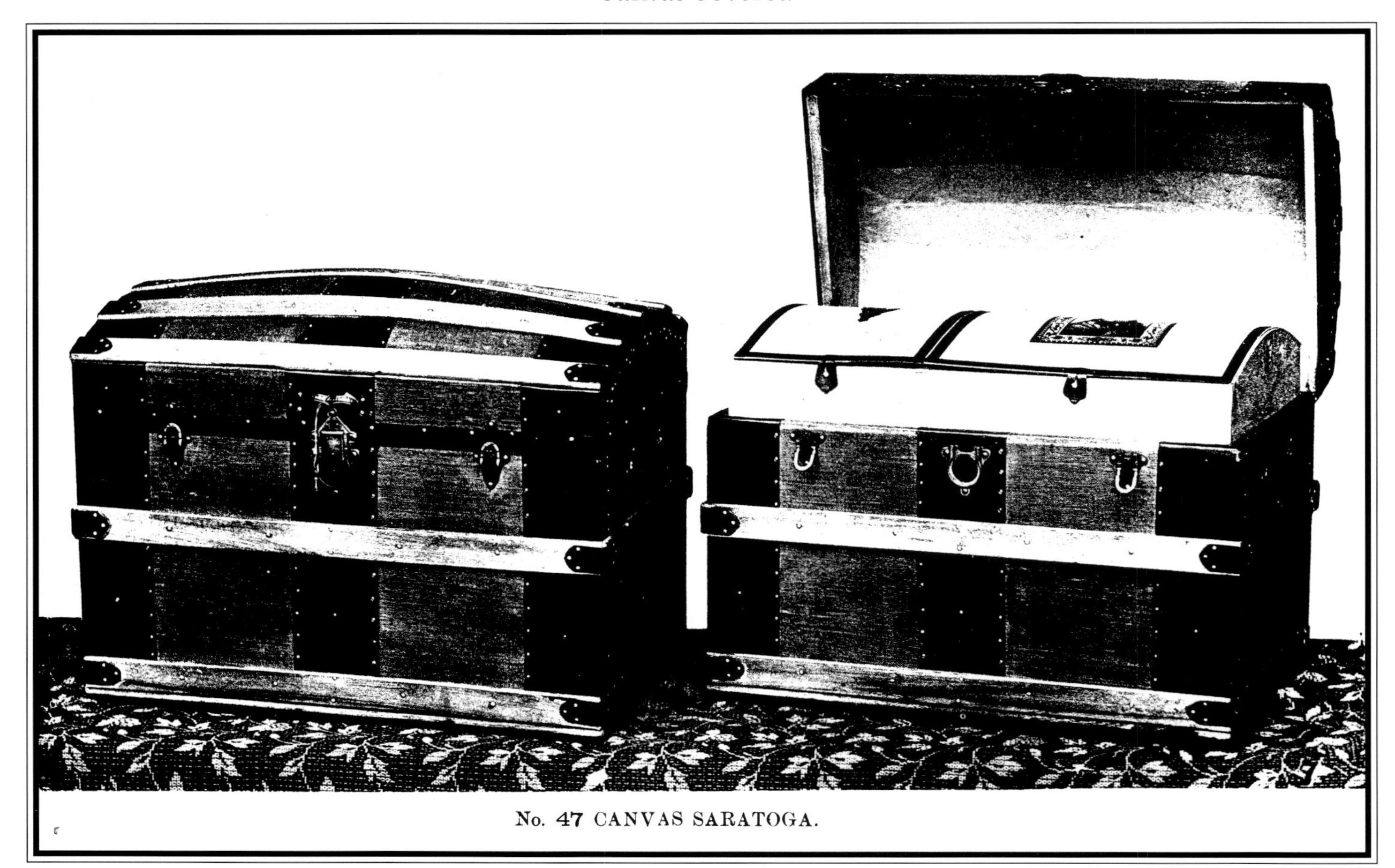

No. 47 CANVAS SARATOGA.

No. 2 CANVAS DRESS.

Paper Lined, one Plain Tray.

No. 3 CANVAS DRESS, Muslin Lined.

Apartments in upper Tray for Hat, &c., and Dress Tray.

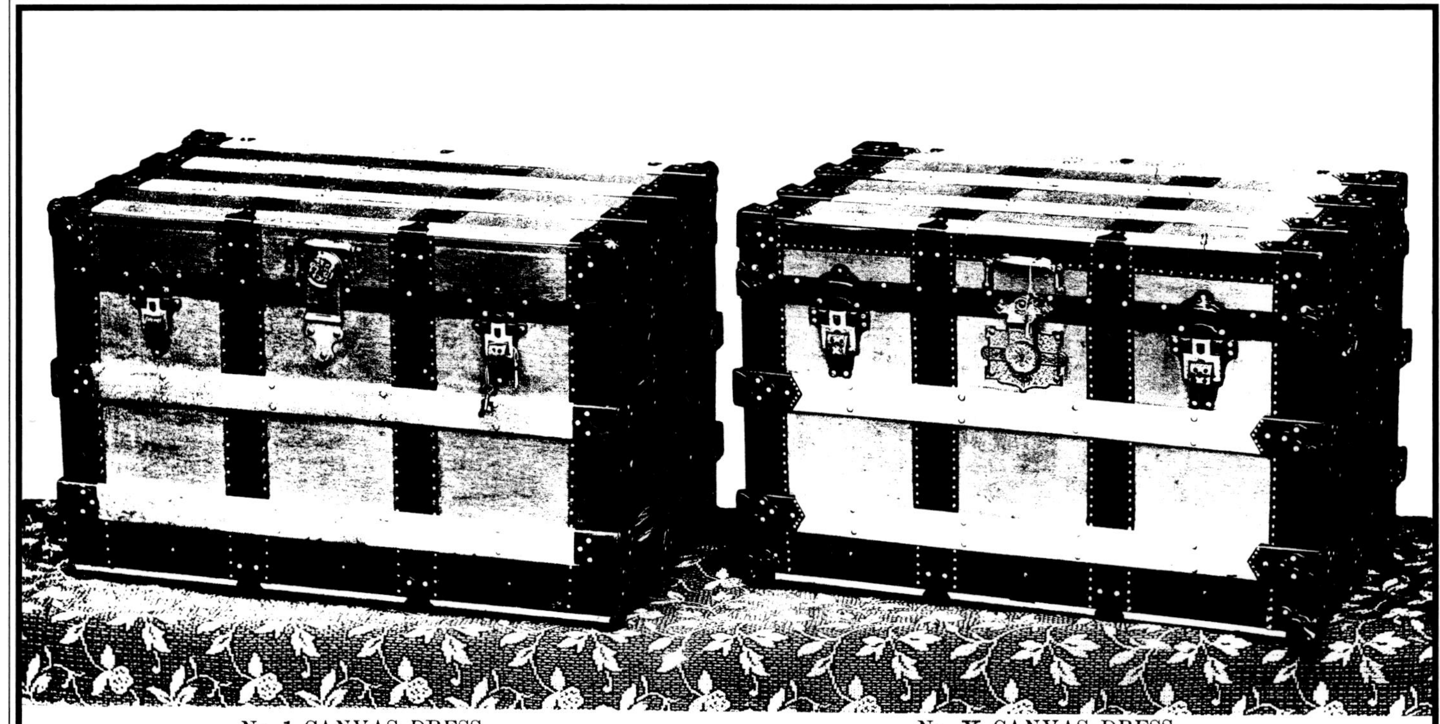

No. 1 CANVAS DRESS.

Muslin Lined.

No. X CANVAS DRESS.

Linen Lined.

Apartments in upper Tray for Hat, &c., and web bottom Dress Tray.

No. 4 CANVAS DRESS, Muslin Lined. No. 5 CANVAS DRESS, Linen Lined.

Leather Bound, Brass Clamps, apartments in upper Tray for Hat, &c., and web bottom Dress Tray.

No. 3 CANVAS STEAMER, Leather Bound.
Brass Clamps, Linen Lined.

No. 2 CANVAS STEAMER, Muslin Lined.

WATER PROOF CANVAS STEAMER, Leather Bound.
Muslin Lined.

No. 1 CANVAS STEAMER, Leather Bound.
Muslin Lined.

$5.45 CANVAS COVERED WALL TRUNK

For better Wall Trunks see Nos. 33K1110 and 33K1116

No. 33K1112 This strong and serviceable Canvas Covered Wall Trunk at a remarkably low price. Cover of trunk is fitted to the body, as shown, in such a manner that cover can be raised or lowered when trunk is set close against wall, whereas with the ordinary trunk this is impossible. Utilized to the greatest advantage when not traveling, this is a particularly convenient style trunk, but we suggest that you consider the many additional advantages of our dresser trunks (which are also wall trunks) namely, Nos. 33K1110 and 33K1116, illustrated and described below, as they are the most convenient trunks ever made. But if you like No. 33K1112, the trunk here described, we assure you that this trunk will give satisfaction, as it is made on strong large basswood box covered with heavy painted canvas, has iron bottom and is bound at ends with black enamel sheet iron. Four strong hardwood slats on top and two around body, as illustrated, brassed steel corner bumpers and clamps, strong rollers, hinges and brass Monitor lock, leather handles, deep set up tray with two compartments separately covered with folding lids and full cloth faced. A wonderful trunk at the price and is guaranteed to give satisfaction. **State size wanted.**

REDUCED PRICES:

Length	Width	Height	Weight	Price
Length, 32 in.	Width, 18½ in.	Height, 19½ in.	Weight, 50 lbs.	Price..$5.45
Length, 36 in.	Width, 23¾ in.	Height, 21¾ in.	Weight, 57 lbs.	Price.. 5.95

$6.85 WONDERFUL CANVAS COVERED TRUNK

Our best medium priced trunk.

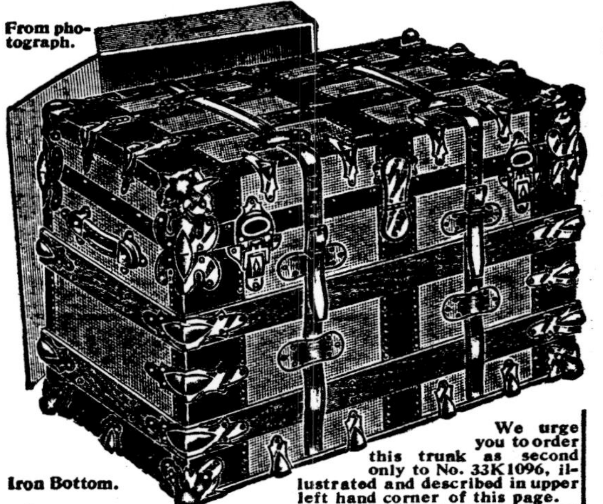

We urge you to order this trunk as second only to No. 33K1096, illustrated and described in upper left hand corner of this page.

No. 33K1070 The equal of this trunk cannot be bought at retail at twice our remarkable price. Extra large size painted canvas covered trunk. Made with four extra wide and heavy hardwood slats on top, and two body slats running clear around. Heavy sheet iron bottom and binding on ends. Has heaviest and strongest malleable brass clamps and reinforcements at every point. Buckle bar bolts. Brass Excelsior lock. Heavy brass corner bumpers and malleable valance clamps at corners where lid and body meet. Strong rollers, and is reinforced with two extra heavy sole leather straps. Contains large, roomy tray with hat box and large side compartment separately covered. A trunk that has been thoroughly tried by us for several years and has never disappointed. We will replace any trunk which does not give satisfaction. When ordering please state size.

Length	Width	Height	Weight	Special price
Length, 32 in.	Width, 21 in.	Height, 23½ in.	Weight, 50 lbs.	Special price..$6.85
Length, 36 in.	Width, 22½ in.	Height, 25 in.	Weight, 60 lbs.	Special price.. 7.55
Length, 38 in.	Width, 23 in.	Height, 25½ in.	Weight, 65 lbs.	Special price.. 7.90
Length, 40 in.	Width, 24 in.	Height, 26 in.	Weight, 72 lbs.	Special price.. 8.25

$5.65 CANVAS COVERED STEEL BOUND TRUNK

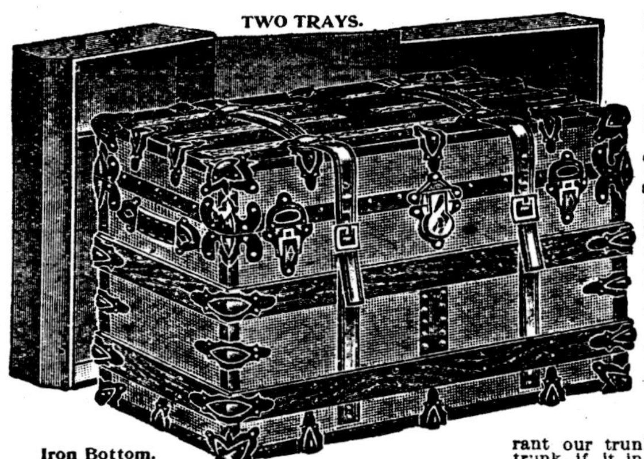

No. 33K1058 Large, Full Size Trunk, thick basswood box, covered with heavy waterproof canvas, entirely bound with japanned angle steel binding, heavy hardwood slats, heavy malleable iron japanned valance clamps, buckle bar bolts, corner bumpers, steel strip clamps, knees, center band and iron bottom, brass Monitor lock, sole leather straps. Contains roomy hinged upper tray with hat box and side compartments separately covered with folding lids and extra skirt tray below. What can you get in the ordinary retail way that will compare in construction, reinforcements and appearance with this trunk? We warrant our trunks absolutely, and this trunk, if it in any way at any time should prove unsatisfactory, we will replace without charge. Please bear in mind the fact that these trunks are built on honor and that we are not afraid to warrant them absolutely. **Remember to give the size wanted and catalogue number when ordering.**

Length, 30 in.	Width, 17½ in.	Height, 19½ in.	Weight, 51 lbs.	Price....$5.65
Length, 32 in.	Width, 18½ in.	Height, 20½ in.	Weight, 56 lbs.	Price.... 6.00
Length, 36 in.	Width, 20½ in.	Height, 22½ in.	Weight, 65 lbs.	Price.... 6.70
Length, 38 in.	Width, 21½ in.	Height, 23½ in.	Weight, 69 lbs.	Price.... 6.95

$6.05 WAGON OR STEAMER TRUNK. CANVAS COVERED. Prices Reduced.

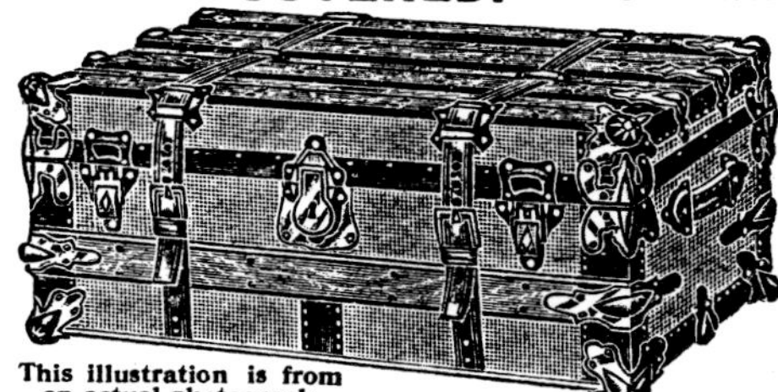

This illustration is from an actual photograph.

No. 33K1136 Painted Canvas Covered Wagon or Steamer Trunk. Four heavy hardwood slats running lengthwise across top, one heavy hardwood slat across front and back, all protected with heavy brass clamps and fancy corner bumpers, brass lock and side bolts. Heavy brass valance clamps at corners where cover meets body. Tray with two compartments separately covered muslin lined throughout. Reinforced as illustrated and strengthened with heavy sole leather straps. Our guarantee: We will replace any trunk of this number which does not prove satisfactory. **State size and catalogue number when ordering a trunk.** Following are **REDUCED PRICES:**

Length	Width	Height	Weight	Reduced Prices
32 inches	19 inches	13½ inches	36 pounds	...$6.05
36 inches	20 inches	13½ inches	40 pounds	... 6.70
38 inches	21 inches	13½ inches	43 pounds	... 7.25

WATERPROOF DUCK COVERED TRUNK

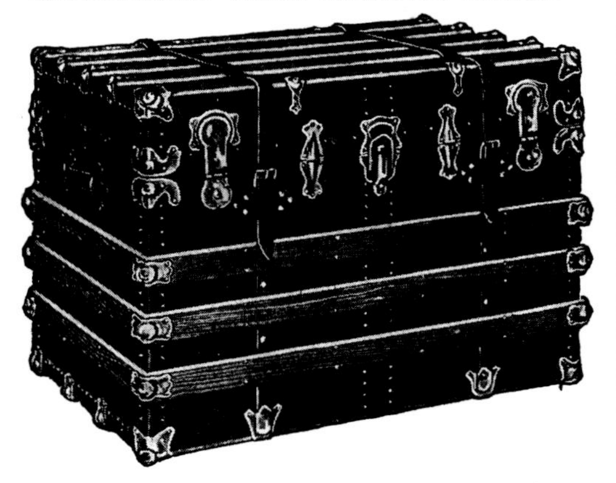

Made of best grade of basswood. Covering water-proof painted duck. Binding and band vulcanized fibre. Trimmings, brassed capital. Slats, hardwood, five on top and three around body. Heavy leather straps. Two Excelsior locks. Deep tray, also dress tray. Neatly lined.

	32-inch	36-inch	40-inch
No. **166** Each.......	**$33.60**	**$36.40**	**$39.20**

Steamer size, otherwise same as above.

	32-inch	36-inch	40-inch
No. **176** Each.......	**$32.20**	**$35.00**	**$37.80**

$2.95 LEADER CANVAS COVERED TRUNK

Taken from Photograph.

No. 33K1055 Painted Canvas Covered Trunk, with iron bottom, sheet iron bound edges, four hardwood slats on top, two slats on sides and ends, steel clamps, knees and corner bumpers top and bottom. Monitor lock, patent bolts, hinges, catches, rollers, etc., leather handles, deep tray and hat box covered. Large size box, paper lined. A low priced trunk which for service is unequaled. While this trunk for the money is guaranteed to be better than can be bought anywhere else, we strongly recommend the purchase of better trunks, as a good trunk will last indefinitely. If interested in a better trunk, we recommend to you our Nos 33K1058 or 33K1070, both illustrated and described on the following page. Remember, though, that at the price we defy anyone to equal the trunk here illustrated and described and which is guaranteed to give satisfaction.

Iron Bottom.

State the size wanted and give catalogue number.

Length, 28 in.	Width, 16¾ in.	Height, 17¾ in.	Weight, 38 lbs.	Special price. **$2.95**
Length, 32 in.	Width, 18¾ in.	Height, 19¾ in.	Weight, 50 lbs.	Special price. **3.65**
Length, 36 in.	Width, 20¾ in.	Height, 21¾ in.	Weight, 57 lbs.	Special price. **4.35**
Length, 38 in.	Width, 21¾ in.	Height, 22¾ in.	Weight, 62 lbs.	Special price. **4.70**

WATERPROOF DUCK COVERED TRUNK

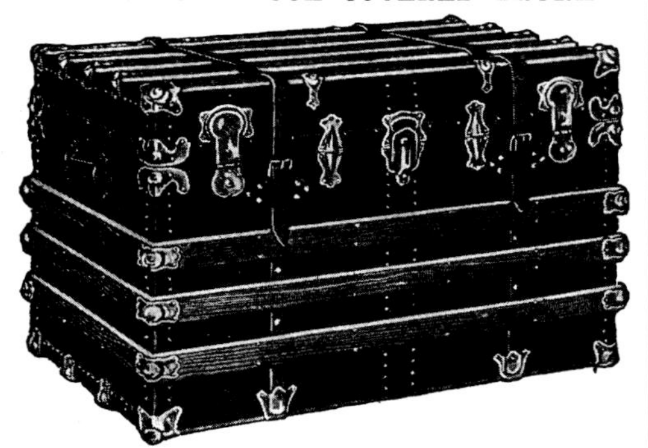

Made of best grade of basswood. Covering water-proof painted duck. Binding and band vulcanized fibre. Trimmings, brassed capital. Slats, hardwood, five on top and three around body. Heavy leather straps. Two Excelsior locks. Deep tray, also dress tray. Neatly lined.

	32-inch	36-inch	40-inch
No. **166** Each	**$33.60**	**$36.40**	**$39.20**

Steamer size, otherwise same as above.

	32-inch	36-inch	40-inch
No. **176** Each	**$32.20**	**$35.00**	**$37.80**

COVERED WATERPROOF CANVAS TRUNK

Heavy Leather Straps

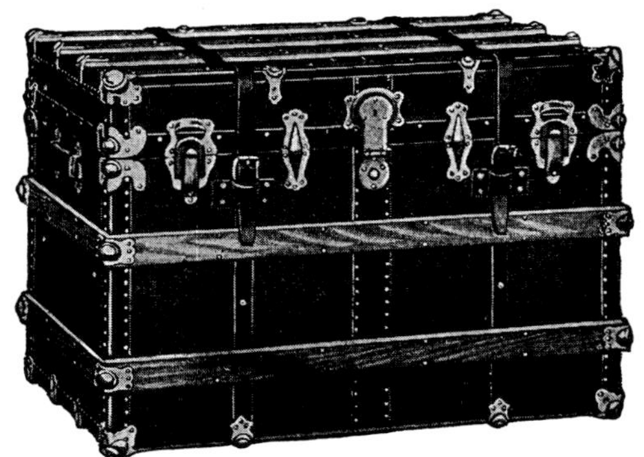

Box, basswood. Covering, waterproof painted canvas. Binding and band, maroon enameled sheet steel. Trimmings, brassed capital. Slats, hardwood. Heavy leather straps. Excelsior lock. Deep tray with full covered lid. Neatly lined.

	32-inch	36-inch	40-inch
No. **106** Each	**$23.80**	**$25.20**	**$26.60**

Steamer size, otherwise same as above.

	32-inch	36-inch	40-inch
No. **116** Each	**$23.10**	**$24.50**	**26.15**

WATERPROOF CANVAS TRUNK

Hardwood Slats

Box, basswood. Covering, waterproofed painted canvas. Binding and bands, vulcanized fibre. Trimmings, brassed capital. Slats hardwood. Heavy leather straps. Excelsior lock. Deep tray with full covered lid. Neatly lined.

	32-inch	36-inch	40-inch
No. **146**	**$26.60**	**$26.95**	**$29.40**

Steamer size, otherwise same as above.

	32-inch	36-inch	40-inch
No. **156**	**$26.15**	**$27.30**	**$28.70**

Leather Covered

No. 2 SOLE LEATHER.

SIZES, **28 in., 30 in., 32 in., 34 in., 36 in.**

Leather corner binding rivetted, heavy straps, steel springs in top, inside cloth lined, tray and top facing imitation leather, heavy duck cover, leather bound.

No. 1 SOLE LEATHER.

SIZES, 30 in., 32 in., 34 in., 36 in.

Heavy russet leather, rivetted French edge, heavy straps, steel springs in top, hickory bottom cleats, linen lined in body and tray, tray and top leather faced, extra duck cover leather bound with corners. A very superior made trunk.

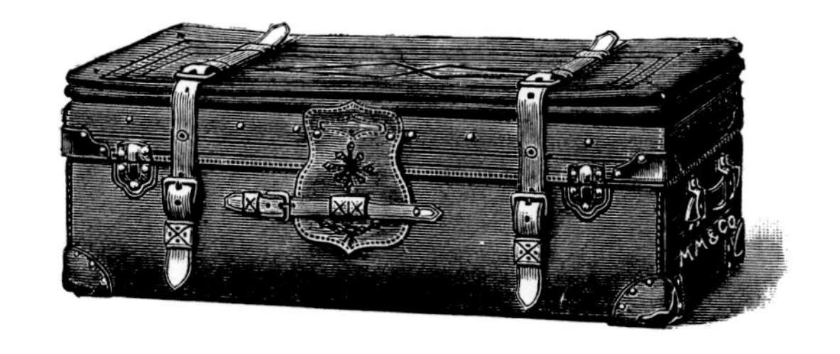

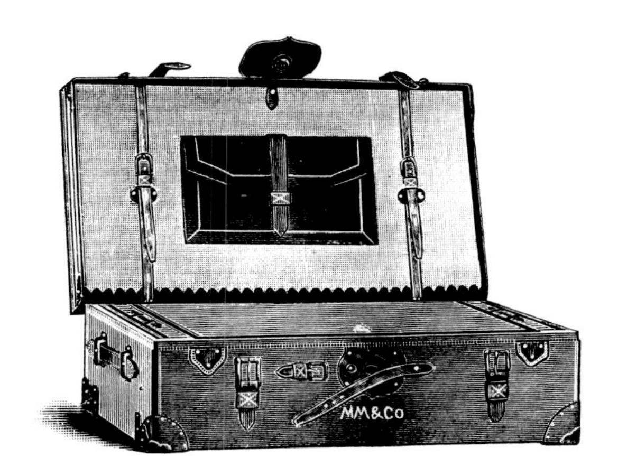

SOLE LEATHER STEAMER TRUNK.

SIZES, 28 in., 30 in., 32 in., 34 in., 36 in.

Made of best quality oak tanned leather, extension bellows top, inside lined with Irish linen throughout, leather pocket on lid.

No. 4 LEATHER SARATOGA.

Muslin Faced—Trays like No. 3.

No. 3 LEATHER SARATOGA.

Muslin Lined throughout.

No. 1 LEATHER GENTS.

Muslin Lined throughout—Trays like No. 2.

GENTS CANVAS "GEM."

Linen Lined throughout—Sargent & Greenleaf Locks

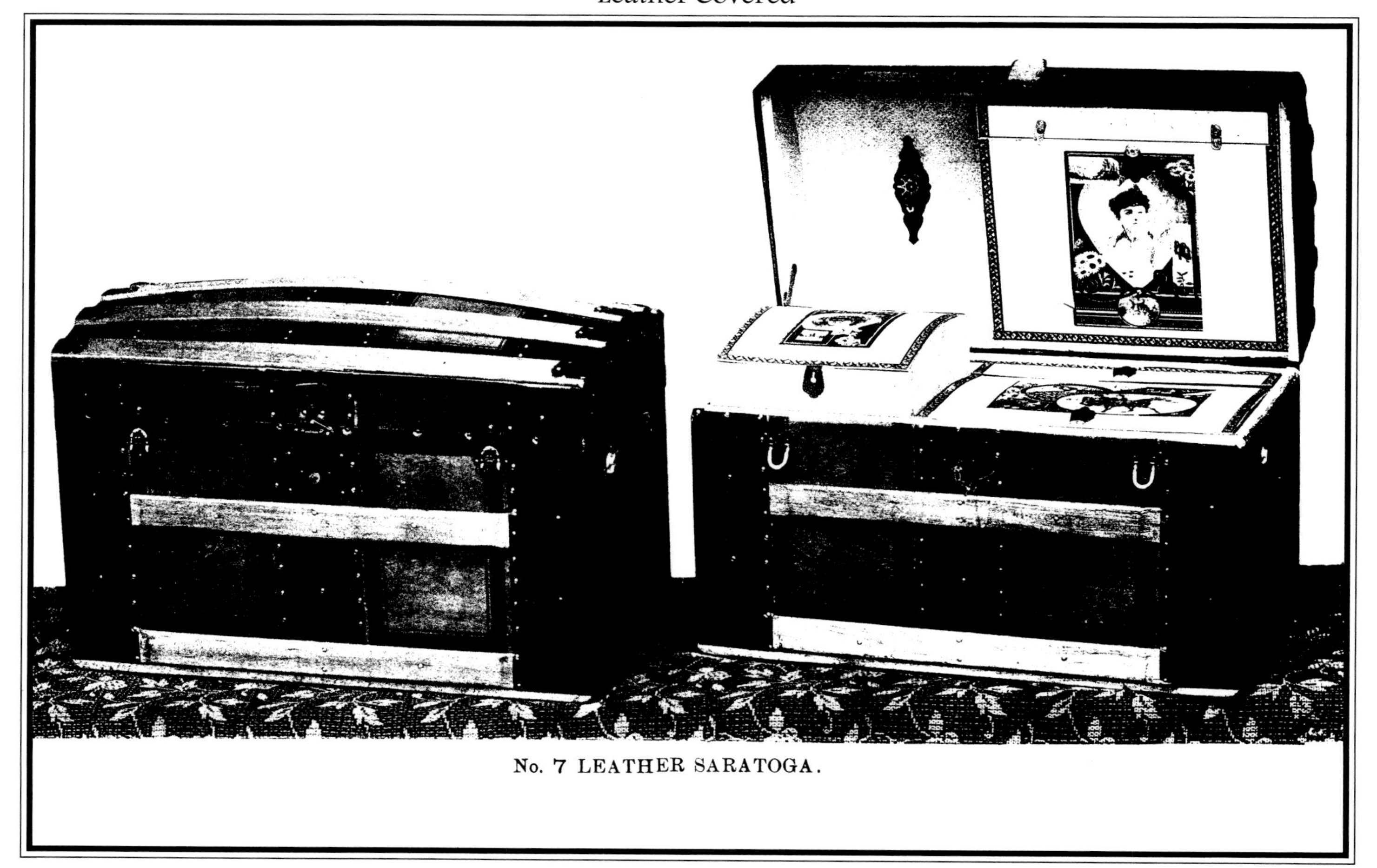

No. 7 LEATHER SARATOGA.

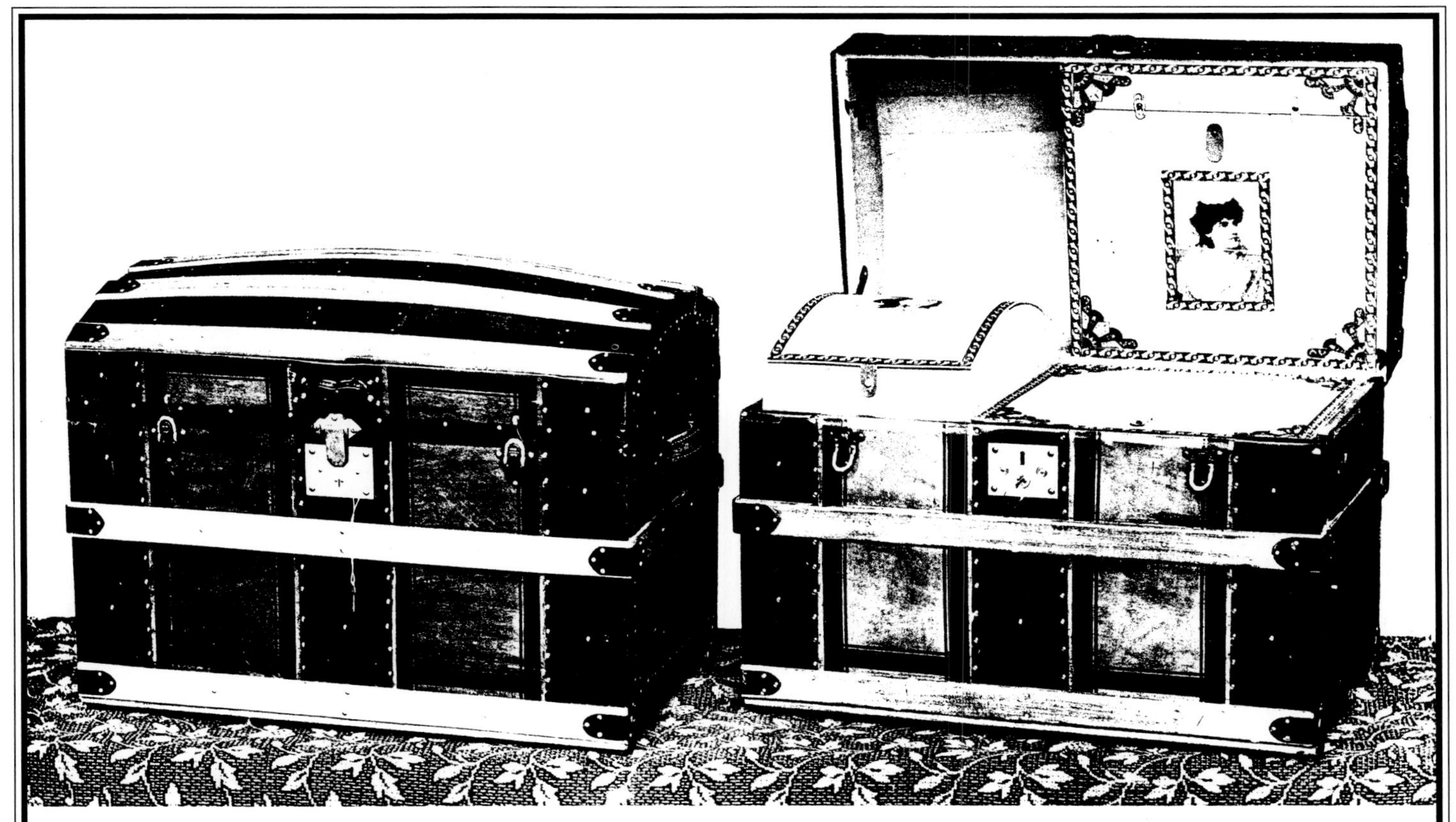

No. 46 LEATHER SARATOGA.

Trays like No. 56.

No. 56 LEATHER SARATOGA.

Muslin Lined throughout.

No. 24 CRYSTAL SARATOGA.

No. 34 LEATHER SARATOGA.

Muslin Faced.

Trays like No. 3.

No. 1 LEATHER SARATOGA, with Dress Tray.

Muslin Lined throughout.

No. X LEATHER SARATOGA, with Dress Tray.

Linen Lined throughout.

Trays like No. 2, with movable Hat Box.

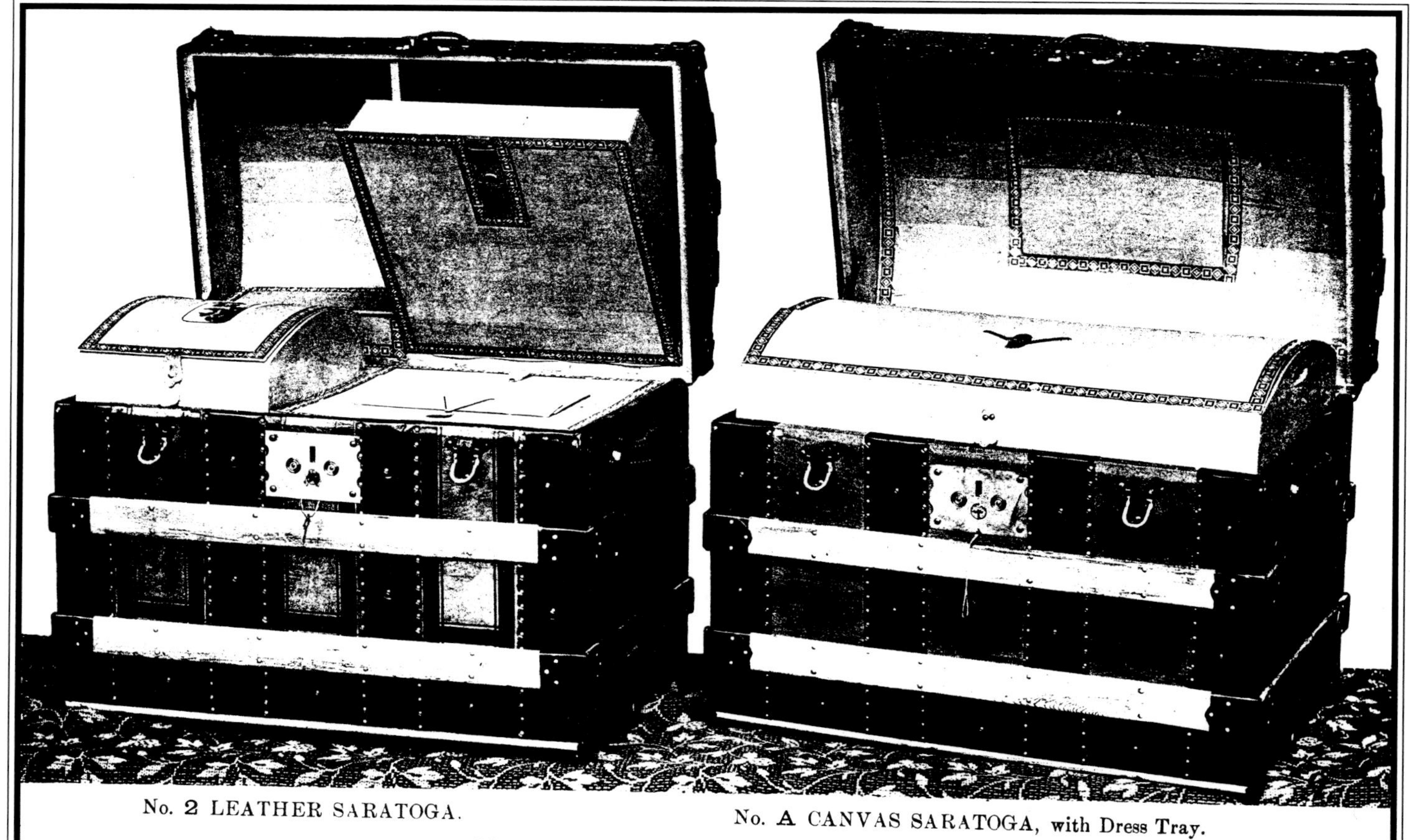

No. 2 LEATHER SARATOGA.

No. A CANVAS SARATOGA, with Dress Tray.

Muslin Lined throughout.

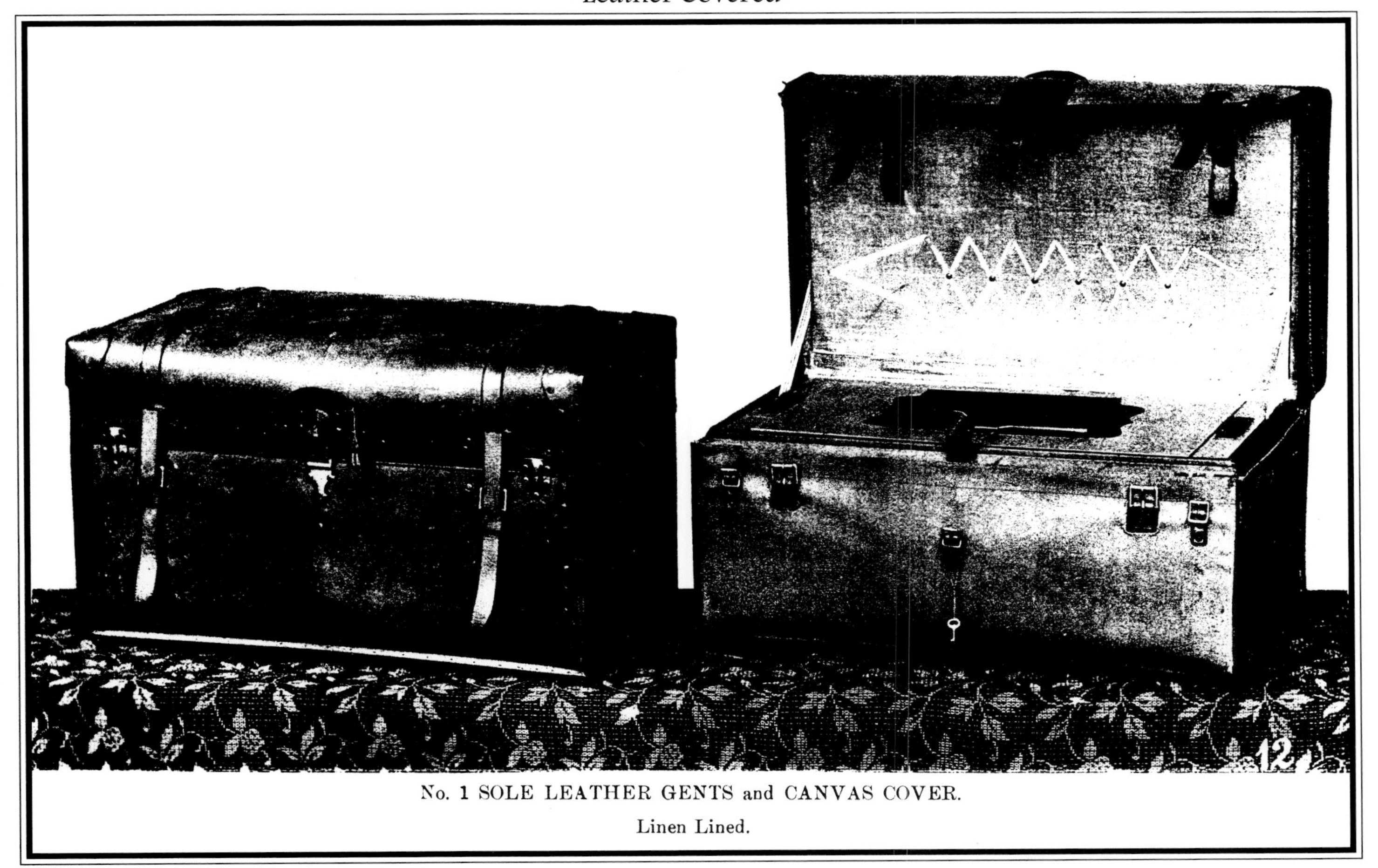

No. 1 SOLE LEATHER GENTS and CANVAS COVER.

Linen Lined.

No. 1 SOLE LEATHER SARATOGA and CANVAS COVER.

Linen Lined. Apartments in upper Tray for Hat, &c., and Web bottom Dress Tray.

WE CALL PARTICULAR ATTENTION TO OUR RAW HIDE BOUND SAMPLE TRUNK

Raw Hide Binding with Wrought Steel Corners.

Everything Riveted.

NO WOOD SLATS

NO NAILS

The most Economical Sample Trunk ever made.

All regular sizes in stock.

This Trunk is in constant use in Clothing, Hat, Dry Goods and Wall Paper trades, etc.

688 BROADWAY
Below Fourth Street

161 BROADWAY
Below Cortlandt Street

723 SIXTH AVENUE
Below 42d Street

NEW YORK

We also make the Largest and Finest Lines of Trunks and Dress Suit Cases, etc., for Family Use
Send for Price List, care of Station A, New York

$11.85 THE STRONGEST TRUNK MADE

Buy this trunk on our recommendation.

TWO TRAYS.

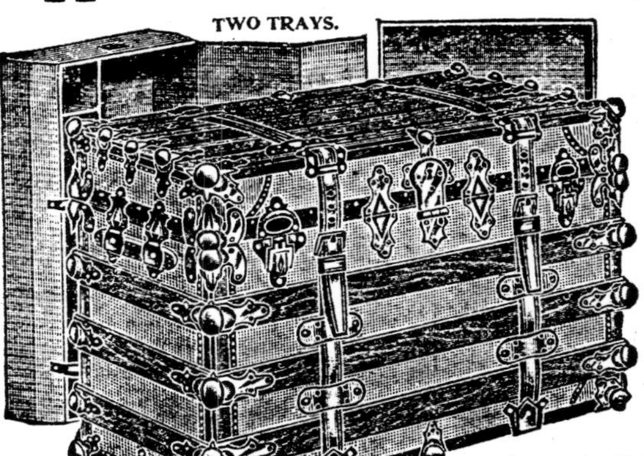

Strongest trunk ever made.

No. 33K1096 Wonderful Leather and Iron Bound Trunk. Built like a battleship because reinforced at every danger point. Extra large and thick basswood box covered with heaviest painted canvas. Flat top, five painted hardwood slats running lengthwise on top, and three heavy hardwood slats running around body of trunk. Front and back angle steel binding, edges heavy cowhide leather bound with fancy leather quarter rounds in corners, heavy leather handles and two heavy cowhide leather straps running through fancy metal and leather loops and tips. Heavy dome set brass plated clamps, knees, corner bumpers, valance clamps, etc. Brass Excelsior lock, patent lifter bolts, socket dowel clamps on front and ends. Heavy hinges, patent rollers, iron bottom, and contains hinged tray with hat box and other compartments, each with folding lid. Additional dress tray, full cloth lined, which fits underneath main tray. You cannot fail to note the thorough manner in which this trunk is made and reinforced. We say, "built like a battleship," and the trunk is sufficient proof of the statement. A great big trunk, which can never be broken in ordinary service. We will replace any trunk so broken. **Do not fail to state the size wanted and catalogue number.**

Length, 32 in.	Width, 18½ in.	Height, 21 in.	Weight, 65 lbs.	**Reduced** price .. $11.85
Length, 34 in.	Width, 19½ in.	Height, 22 in.	Weight, 69 lbs.	**Reduced** price .. 12.60
Length, 38 in.	Width, 21½ in.	Height, 24 in.	Weight, 82 lbs.	**Reduced** price .. 14.10
Length, 40 in.	Width, 22½ in.	Height, 25 in.	Weight, 92 lbs.	**Reduced** price .. 14.85

Kartavert

GENTS' KARTAVERT.

SIZES, 28 in., 30 in., 32 in., 34 in.

This trunk is made of the very lightest and most substantial material used for making trunks, far superior to sole leather for service. The entire trunk is of Kartavert, the two center bands, the binding on end edges and edges front and back, top and bottom, are all of heavy Kartavert, and all firmly secured to the trunk with brass rivets, in keeping with the solid cast brass trimmings of our own design, made for this trunk and thoroughly secured by solid rivets. Strong corner and body clamps, box shoulder and valance clamps, front and back, heavy Taylor bolts, heavy brass Excelsior lock of our own design, double backed strap hinges, heavy leather sliding handles. Straps in linen compartments and dress tray. Padded silk hat rest. Trunk lined with Irish linen throughout. See page 51.

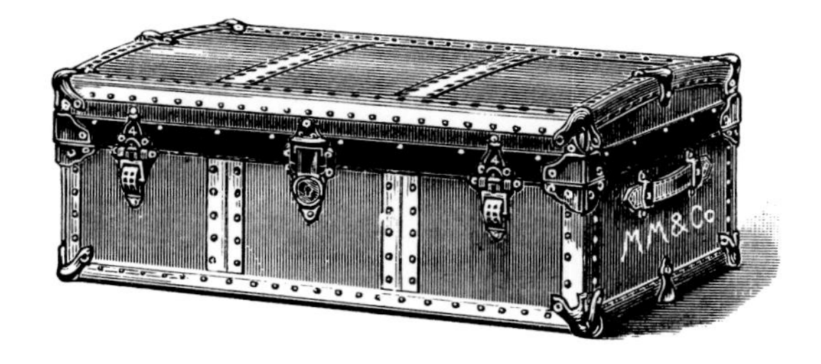

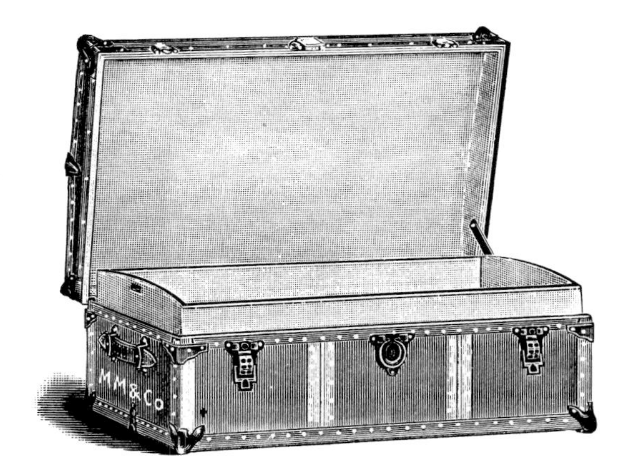

KARTAVERT STEAMER.

SIZES, 30 in., 32 in., 34 in., 36 in.

Made of Kartavert, the most substantial material used for making trunks. Binding and bands are of very heavy Kartavert. All the trimmings are solid cast brass of our own design and securely fastened with solid rivets and burrs. Heavy Excelsior lock of our design. Binding and bands all securely riveted to the trunk. Inside lined with Irish linen.

KARTAVERT TRUNKS.

The Best Sample Trunk Made.

For Dry Goods, Clothing, Blankets, Robes, Knit Goods, Etc., Etc.

STOCK SIZES, 30 in., 32 in., 34 in., 36 in., 38 in.

For Boots and Shoes, Stock Size, 36 inches.

Special harness trunks in stock.

Trunks to suit any business made special sizes to order.

Interior cases telescope style, either of Kartavert, common red fibre or canvas covered, for boot and shoe or other lines of goods made to order. Canvas covered shoe cases in stock.

Bargains in Wardrobe Trunks, Hat Cases and Boston Bags AT FACTORY PRICES

Workmanship and Finish Unsurpassed. Positively the Greatest Values Possible at Prices Quoted.

FULL SIZE RAWHIDE BOUND COMBINATION WARDROBE

The highest grade wardrobe made. America's De Luxe wardrobe for the man or lady that wants the best money can buy. Has every improvement and convenience conceivable.

Black vulcanize fibre covered. Rawhide binding and two rawhide center bands. Box of 3-ply veneer, fibre lined. Studded with saddle nails. Brass plated steel trimmings. Solid brass paracentric lock. Open top, velvet lined. Purple moire keratol faced. Cloth lined throughout. Five drawers for linen, hats, etc. Top drawer divided into three compartments. Secret drawer for jewelry. Ladies' and gentleman's hat drawers. Locking device which locks all drawers. Dust proof door, laundry bag, umbrella strap, shoe pockets. Patented clothes retainer. Patented steel frame. Size: height, 45 inches; width, 24½ inches; depth, 22 inches.

No. W791 Each..$205.65

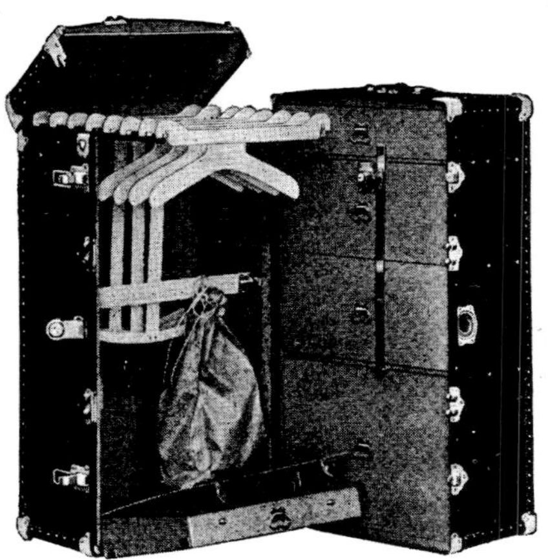

FULL SIZE OPEN TOP VULCANIZED FIBRE WARDROBE

Elegant value in this popular Wardrobe.

Black vulcanized fibre covering and binding. All edges rounded. Three ply veneer box fibre lined. Brass plated steel trimmings. Spring lock and draw bolts. Fancy cloth lined. Open top velvet lined. Shallow top drawer with divisions. Bottom drawer extra deep for Hat box. Drawer locking device. Shoe box. Laundry bag. Full set of garment hangers. Ratchet bar follower. Height, 42 inches; width, 23½ inches; depth, 22 inches.

No. W733 Each..$91.00

FULL SIZE OPEN TOP ROUND EDGE FIBRE COVERED WARDROBE

An exceptional high grade wardrobe that is remarkably well made, compact and has every patented convenience conceivable.

Black vulcanized fibre covering and walnut fibre binding. Box is made of three-ply veneer, fibre lined. Brass plated steel trimmings. Brass Excelsior lock and draw bolts. Fancy cretonne faced. Full cloth lined. Five drawers divided for linen, hats, etc. Top drawer divided into 3 compartments. Hat crown in bottom drawer. Laundry bag and shoe pockets. Patented trolley and clothes retainer. Patented steel frame. Bar drawer locking device, locks all drawers at one time. Size: height, 45 inches; width, 23½ inches; depth, 22 inches.

No. W727 Each..$125.15

A High Grade Line of Dress, Steamer and Wardrobe Trunks

They Are Solidly Constructed. Built to Give Satisfaction and Are Fully Warranted. These Wardrobe Trunks Are Full Size and Can Be Used For Gent's or Ladies' Garments. A Very Serviceable Outfit.

IT IS OUR AIM

To handle Merchandise of dependable quality. It gives satisfaction to everybody, it makes a permanent and satisfied customer which is the best asset a merchant can have, as your business will then

GROW AND PROSPER

OPEN TOP WARDROBE TRUNK

Will be a real pleasure to you when you travel. It has our sturdy veneer wood box, covered and lined with hard vulcanized fibre, lined with a beautiful cretonne, divided shoe pocket and hangers for ladies' or gentlemen's garments, five roomy drawers for shirts, etc., and our protected convertible hat drawer, which can be turned over to accommodate large hats for ladies or used as a regular drawer. This is the most convenient style of wardrobe made for ladies that like to travel in comfort. Height, 40 inches; 21½ inches wide and 22 inches deep. Heavy brass snap lock and bolts.

No. **606** Each..**$89.60**

WARDROBE TRUNK

Three-quarter size—42x21½x18—and is especially suited for gentlemen or ladies desiring to take along only eight or ten suits or dresses. It has a convertible drawer for carrying ladies' large hats that can be used for a regular drawer if desired, and four regular drawers for underwear, etc. Has a divided shoe pocket; neat cretonne lining; good lock and bolts and heavy dowels. The construction is the best 3-ply veneer box, covered and lined with hard vulcanized fibre.

No. **556** Each..**$70.00**

Same general construction as above, in the full size—40x21½x22—and will accommodate several more garments. Remember the construction, **5-ply veneer wood, hard vulcanized fibre inside and outside.** High grade hardware.

No. **566** Each..**$76.40**

FULL SIZE FIBRE COVERED WARDROBE
No. W739 (Description below), each....................$71.50

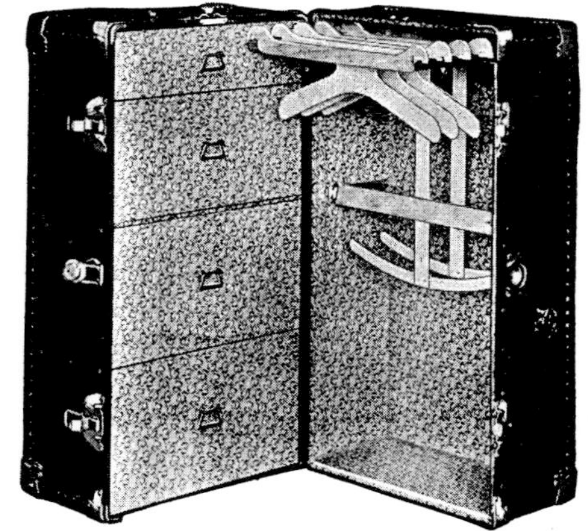

ROUND EDGE FULL SIZE WARDROBE
No. W703 (Description below), each....................$53.35

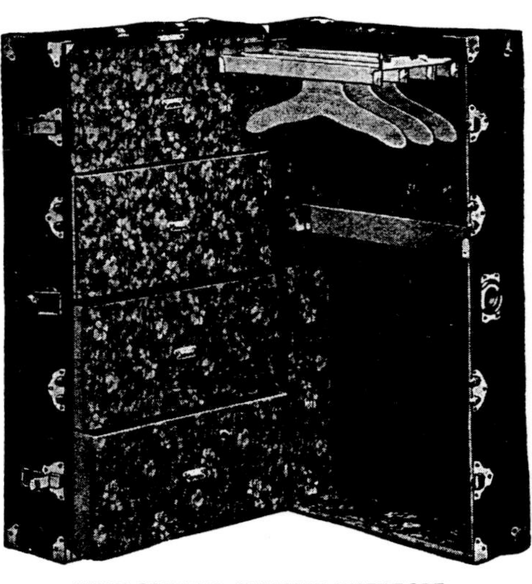

FIBRE COVERED STEAMER WARDROBE
No. W709 (Description below), each....................$50.70

WARDROBE TRUNK

Its capacity is surprising. Small enough to carry in a taxi, to keep open in a steamer stateroom and easily stored in a small space. Round edge fibre covered wardrobe trunk. Brass plated trimmings. Excelsior lock. Cloth lining. Three drawers for linens, hats, shoes, etc. Double trolley holding four 3-ply garment hangers. Size—Height, 36 inches; width, 14 inches; depth, 22 inches.

No. 7156 Each.......................**$52.65**

WARDROBE TRUNK

Vulcanized fibre covering and binding. Brass plated spring lock and steel trimmings, cretonne lining, drawers for linen, hats, etc. Double trolley. Patented self-locking clothes retainer. Patented steel frame. Size—Height, 45 inches; width, 23½ inches; depth, 21½ in.

No. **7426** Each.......................**$95.75**

WARDROBE TRUNK

Full size combination trunk for men and women. Covered with dark green vulcanized fibre; bound with darker green fibre. Three-ply veneer box, rounded corners. Lock is of the parancentric style. The bolts are the draw style. Dowels, corners, clamps and other trimmings are of heavy steel brass plated. Trunk is faced with cretonne and linen lined. two lower drawers combine to make a hat box 21x20x10½ inches, 4 pairs of shoes. Equipped with double trolley and full set of 3-ply veneer hangers. Clothes retainer is most practical and simplest used in wardrobe trunks.

Height, 45 inches; width, 23½ inches; depth, 22 inches.

No. **7696** Each.......................**$135.90**

$10.95 WALL DRESSER TRUNK

We could offer a cheaper dresser trunk, but you would not want it.

PRICES REDUCED.

Iron Bottom.

Illustration is made from an actual photograph.

No. 33K1116 Special Wall Dresser Trunk, painted canvas covered large thick basswood box, thoroughly reinforced and protected by heaviest malleable japanned trimmings, with No. 4 buckle bar bolts in front and on ends, has brass Excelsior lock, strong patent catches, hinges and rollers. Valance clamps at corners where cover strikes body. The most convenient style trunk ever made. Any part accessible at all times. Upper part has three compartments and body contains three capacious cloth faced drawers, metal bound, resting on steel supports. Exactly as illustrated. One of the most serviceable and convenient of trunks. Retails at $18.00 to $20.00 and at our price is an exceptional value. A favorite with ladies. We cannot afford to misrepresent. Your money's worth here or your money back. **Please state size wanted.**

Following are Reduced Prices:

Length	Width	Height	Weight	Reduced price
32 in.	18½ in.	22 in.	62 lbs.	$10.95
36 in.	20½ in.	24 in.	74 lbs.	12.45
38 in.	21¼ in.	25 in.	80 lbs.	13.20

$17.95 STRONGEST, MOST CONVENIENT BUREAU TRUNK MADE

Recommended as our best.

Accessible at all times.

Taken from photograph.

No. 33K1110 The strongest, heaviest and most convenient Wall Dresser Trunk ever made. Extra large thick basswood box covered with heavy waterproof painted canvas and reinforced and protected with olive enameled steel binding. Brass plated heels, corner bumpers and clamps, No. 4 buckle bar bolts, brass Excelsior lock, strongest and heaviest malleable catches, valance clamps at corners of lid and strong socket dowels joining cover and body, four heavy hinges and rollers, and has heavy stitched leather handles, and is reinforced with two heavy sole leather straps, protected with metal and leather strap loops. The illustration is an exact photographic reproduction of this trunk. Linen lined with genuine Holland linen facing, three strapped pockets and three roomy compartments in lid of trunk and has three large, roomy drawers arranged with movable hat form, and extra compartment in bottom of trunk. For strength and durability, coupled with its many convenient features, this trunk stands without a peer. The strongest guarantee ever put upon a trunk; we will replace any trunk if at any time it is proven unsatisfactory. **Please state size wanted.** Sizes as follows:

FOLLOWING ARE REDUCED PRICES:

Length	Width	Height	Weight	Reduced price
32 in.	20 in.	24 in.	78 lbs.	$17.95
36 in.	22 in.	26 in.	88 lbs.	20.95
38 in.	24 in.	28 in.	98 lbs.	22.45
40 in.	26 in.	30 in.	108 lbs.	23.95

WALL TRUNKS.

This cut shows comparative space from wall taken up by Old Style and ***Peerless Wall Trunk.*** (Difference in space 9 inches.)

What are They?

They are the Most Convenient
Durable and Practical Constructed

BECAUSE:

They can be opened while standing closely against the wall without pulling same forward, doing away with heavy lifting and straining, also avoiding tearing of carpets and mutilating walls.

They have six solid corners; the back solid and have such general protection by peculiar wall trunk attachments and construction as produces beyond question an almost perfect trunk as to convenience, strength and lasting powers.

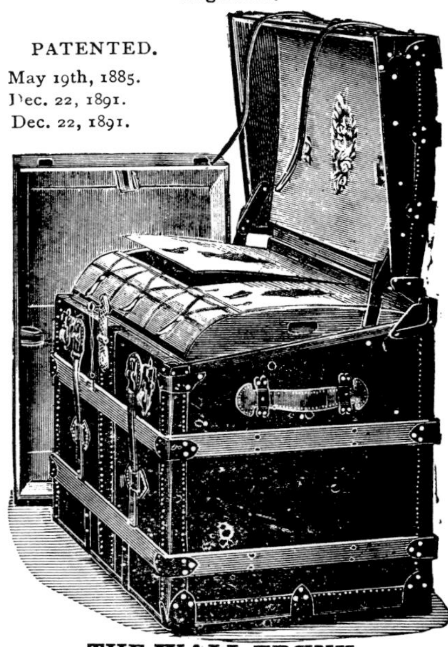

THE WALL TRUNK.

Manufactured Only By and Sold and Controlled

IN

**Indiana, Illinois, Western New York
Ohio, Michigan, and Pennsylvania,**

BY

EGGEMAN, DUGUID & CO.,
TOLEDO, OHIO.

IN

The Western, Southern and Eastern States,

BY

STROMBERG, KRAUS & CO,
LOUISVILLE, KY.

TRAVELING BAGS

GLADSTONES

No. 7 DUCK.

No. 3 DUCK GLADSTONE.

SIZES, 14 in., 16 in., 18 in., 20 in., 22 in.

No. 7 Duck. Cloth lined, nickel trimmings, Jap. frame.

No. 3 Duck Gladstone. Leather corners, cloth lined, pocket on door, Jap. frame.

No. 2 SPLIT PLAIN.

SIZES, 14 in., 16 in., 18 in., 20 in., 22 in., 24 in.

Brown pebbled split leather, Jap. frame, cloth lined, nickel trimmings.

No. 9 DUCK.

No. 2 GLADSTONE.

SIZES, 14 in., 16 in., 18 in., 20 in., 22 in., 24 in.

No. 9 Duck. Cloth lined, nickel trimmings, Jap. frame.

No. 2 Split R. R. Same style as No. 9 Duck. Made of pebbled split leather.

No. 2 Gladstone. Pebbled split leather, cloth lined, Jap. frame.

No. 11 GLADSTONE.

SIZES, 14 in., 16 in., 18 in., 20 in., 22 in.

Imitation alligator leather, cloth lined, nickel trimmings, Jap. frame.

No. HH GLADSTONE.

SIZES, 14 in., 16 in., 18 in., 20 in., 22 in.

Pebbled leather, cloth lined, Jap. frame, nickel trimmings.

No. A X GLADSTONE.

SIZES, 14 in., 16 in., 18 in., 20 in., 22 in.

Brown pebbled grain leather, nickel trimmings, Jap. frame, cloth lined.

No. B B GLADSTONE.

SIZES, 14 in., 16 in., 18 in., 20 in., 22 in.

Pebble grain leather, Jap. frame, brown imitation leather lining, nickel trimmings.

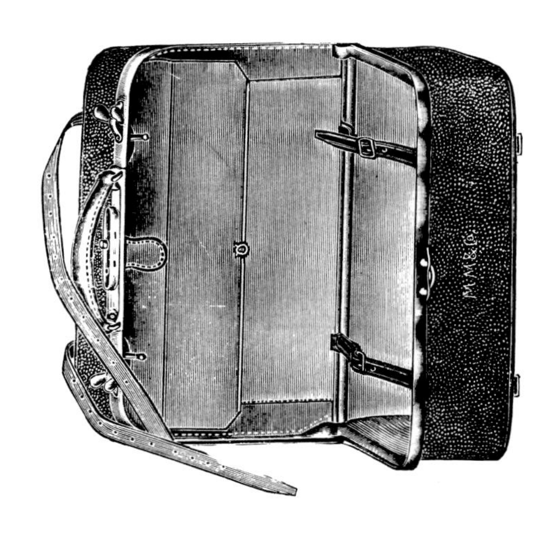

No. D X GLADSTONE.

SIZES, 14 in., 16 in., 18 in., 20 in., 22 in.

Pebbled grain leather, Jap. frame, leather lined, nickel trimmings.

No. R O GLADSTONE.

SIZES, 14 in., 16 in., 18 in., 20 in., 22 in.

Brown pebbled and olive grain leather, covered frame, leather lined, spring catches, nickel corner guards.

No. RE GLADSTONE.

SIZES, 16 in., 18 in., 20 in., 22 in.

Olive and brown pebbled grain leather, covered frame, leather lined, spring catches, nickel corner guards.

No. N N GLADSTONE.

SIZES, 16 in., 18 in., 20 in., 22 in.

Brown pebbled grain leather, pig leather lined, covered inlays, covered frame, covered buckles, shirt pocket on door, English snap catches, shoes on bottom.

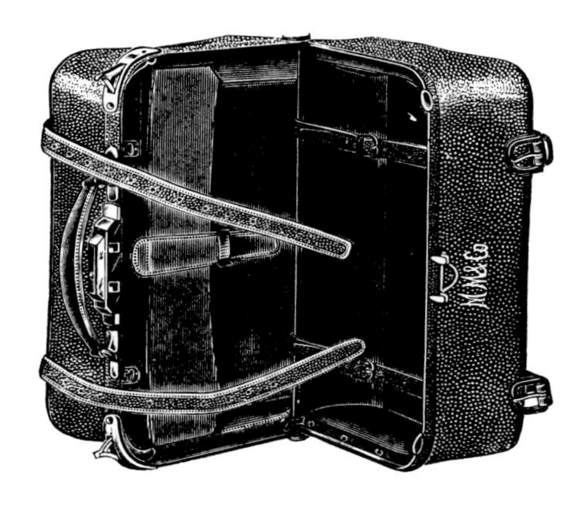

No. R R GLADSTONE.

SIZES, 16 in., 18 in., 20 in., 22 in.

Brown pebbled grain leather, leather lined, covered frame, covered buckles, shirt pocket on door, patent corner catch and guard, shoes.

No. KK GLADSTONE.

SIZES, 16 in., 18 in., 20 in., 22 in.

Grain leather, fancy colors; pig leather lining; shirt pocket on back of door; all straps double and stitched, and with leather covered buckles; double hinged shoe on frame; Columbus handle; brass trimmings.

No. SS DRESS SUIT GLADSTONE.

SIZES, 20 in., 22 in., 24 in.

Made in Fancy Colors, Grain Leather and Alligator.

Pig leather lined; portfolio on door for collars and cuffs; covered frame and inlays; fine nickel trimmings; Columbus handle; patent corner catch; shoes. Most complete gent's bag in the market.

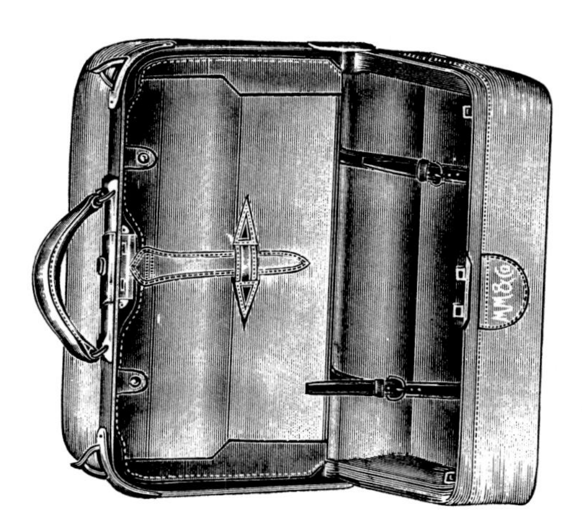

No. N E GLADSTONE.

SIZES, 16 in., 18 in., 20 in., 22 in.

Made in Fancy Colors, Grain and Alligator Leather.

English frame, bag hand sewed to frame; fine leather lined, leather covered buckles on inside straps; Columbus handle; brass trimmed.

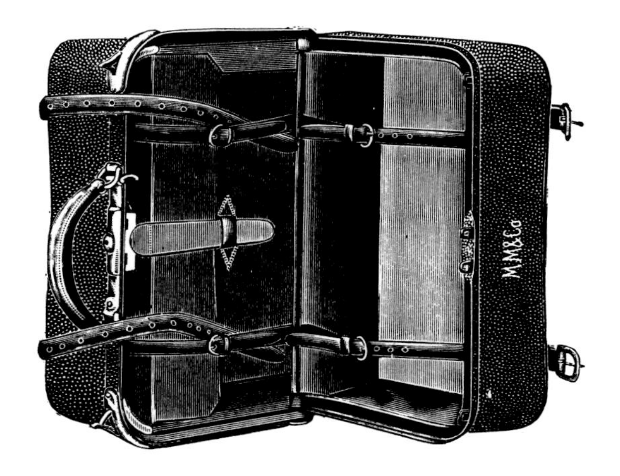

No. 1 ENGLISH GRAIN GLADSTONE.

SIZES, 16 in., 18 in., 20 in., 22 in.,

Made in brown pebbled grain and fancy colors, grain leather; English frame, hand sewed to frame; pig leather lined, shirt pocket, brass trimmings, covered buckles, shoes.

SAFETY GLADSTONE.

SIZES, 16 in., 18 in., 20 in., 22 in.

Pebbled grain leather, heavy steel frame, bag hand sewed to frame. Heavy bolt lock, pig leather lining, shirt pocket, double and stitched straps, leather covered buckles.

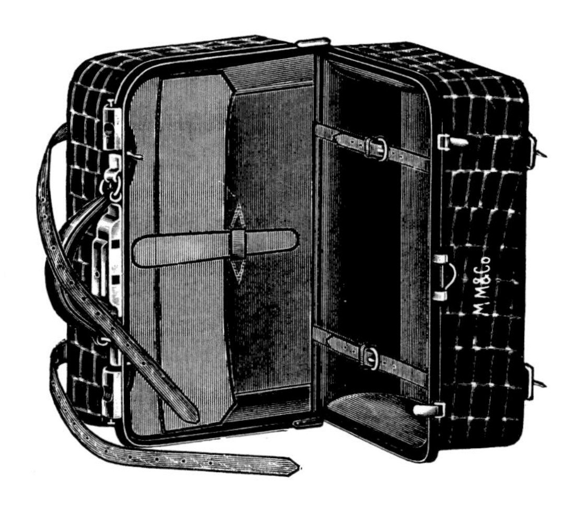

ALLIGATOR GLADSTONE. RIVETTED FRAME.

SIZES, 16 in., 18 in., 20 in., 22 in.

Real alligator, pig leather lining, shirt pocket on back of door; alligator straps, double and stitched; all covered buckles on straps, double hinge shoe on frame.

ENGLISH ALLIGATOR GLADSTONE.

SIZES, 16 in., 18 in., 20 in., 22 in.

Real alligator, pig leather lined, shirt pocket on back of door. All straps double and stitched, leather covered buckles, brass trimmings.

PACIFIC.

SIZES, 14 in., 16 in., 18 in., 20 in., 22 in., 24 in.

Black rubber cloth, patent frame, flat key lock.

The "Improved Gladstone," an entirely new and patented article is being displayed by Krick, Burger & Co., one of the later concerns who have established themselves here. As will be seen by the illustration it contains a bag within a bag, the inner one being designed either for a medicine carrier, toilet case, or shirt holder.

The inner case is readily removed, while the outside grip is fitted with all the most modern devices in straps and catches. They are made in several qualities and shades, those prevailing being russet, seal and olive.

On the whole the styles and qualities of leather goods shown this year are a distinct advance on all previous efforts, and show much more mechanical ingenuity.

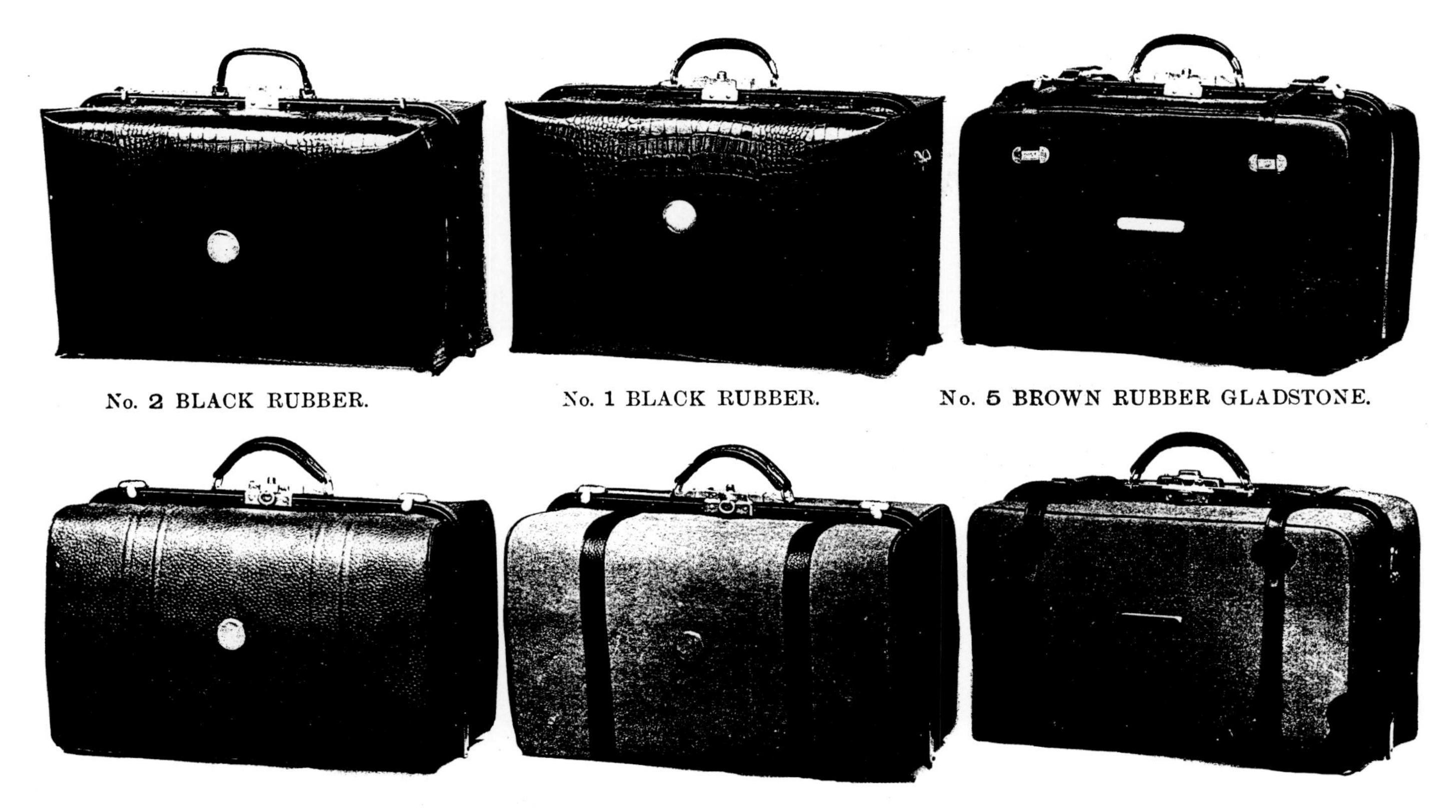

No. 2 BLACK RUBBER.

No. 1 BLACK RUBBER.

No. 5 BROWN RUBBER GLADSTONE.

No. 51 BLACK LEATHER.
No. 52 BROWN LEATHER.

No. 21 CANVAS.

No. 23 CANVAS GLADSTONE.

No. 53 BROWN LEATHER GLADSTONE.

No. 54 DARK LEATHER GLADSTONE. Imitation Alligator.

No. 153 BROWN GRAIN LEATHER GLADSTONE

No. 207 BROWN GRAIN LEATHER GLADSTONE.

No 206 BROWN GRAIN LEATHER GLADSTONE. Linen Lined.

No. 259 BROWN GRAIN LEATHER GLADSTONE. Leather Lined.

No. 255 BROWN GRAIN LEATHER GLADSTONE. Leather Lined.

No. 257 BROWN GRAIN LEATHER GLADSTONE. Leather Lined.

No. 351 BROWN GRAIN LEATHER GLADSTONE. Steel Frame, Linen Lined.

No. 353 BROWN GRAIN LEATHER ENGLISH GLADSTONE.
Brass Trimmings. Leather Lined.

No. 403 LIGHT ALLIGATOR GLADSTONE.
No. 404 DARK " "
Leather Lined.

No. 453 DARK ALLIGATOR ENGLISH GLADSTONE.
No. 454 LIGHT " " "
Brass Trimmings. Leather Lined.

GENUINE COWHIDE LEATHER CASE

Made of genuine cowhide. Cowhide corners. Double riveted, with flat head rivets. Cowhide straps all around. Stitched-on ring handle. Good lock. Linene lined. Shirt pocket.

		24-inch	26-inch
No. **1366**	Brown	**$33.60**	**$35.00**

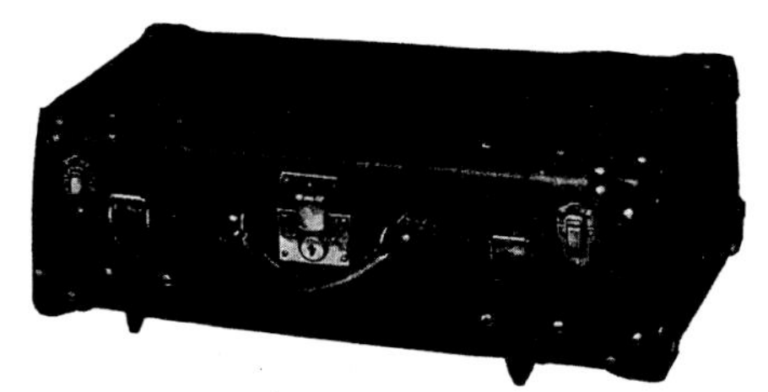

AN EXTRA DEEP GENUINE COWHIDE CASE

Genuine cowhide case. Extra deep. Straps all around reinforced with fancy leather corners riveted on. Good lock and catches. Cloth lining and shirt pocket.

		24-inch	26-inch
No. **1386**	Brown	**$39.75**	**$42.00**

GLADSTONE BAGS

The Most Popular Bag For the Up-to-Date Traveler.

Selected cowhide. A distinctive and convenient bag, divided in the center by partitions and supplied with pockets which keep everything in place. Sewed frame and hand-sewed corner pieces. Solid brass mountings, double post lock and flat catches. Linen or leather lined.

	20-in.	22-in.	24-in.
Brown, linen lined. No. **3026**	**$47.60**	**$50.40**	**$53.20**
Smooth black, linen lined. No. **3036**	**47.60**	**50.40**	**53.20**
Smooth brown, leather lined. No. **3066**	**58.80**	**61.50**	**64.40**
Smooth black, leather lined. No. **3076**	**58.80**	**61.50**	**64.40**

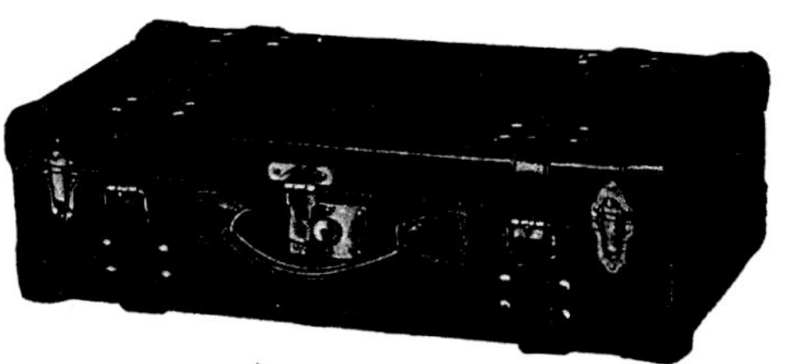

EXTRA DEEP GENUINE COWHIDE CASE

Deep case made of genuine cowhide. Fancy sewed corners. Sewed handle loops with flat leather handle. Side hasp lock and solid brass bolts. Cowhide straps all around case. Genuine linen lined. Small pockets on each end of case. Shirt pocket in lid.

		24-inch	26-inch
No. **1466**	Brown	**$56.80**	**$61.50**

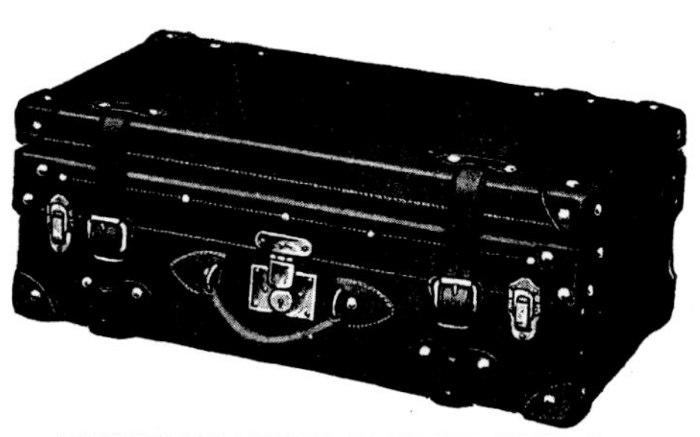

HEAVY LEATHER BELLOW SUIT CASE

Heavy leather in Windsor grain. Heavy cowhide corners. Double bellows. Stitched-on ring handle. Good lock and catches. Heavy cowhide straps all around. Cloth lined and following board.

		24-inch	26-inch
No. **1236**	Each	**$29.40**	**$32.20**

PORTMANTEAU.

SIZES, 20 in., 22 in., 24 in., 26 in.

Fine oak tanned russet leather, Irish linen lined, following board in body, leather pocket on door, leather covered buckles on outside straps, Excelsior lock.

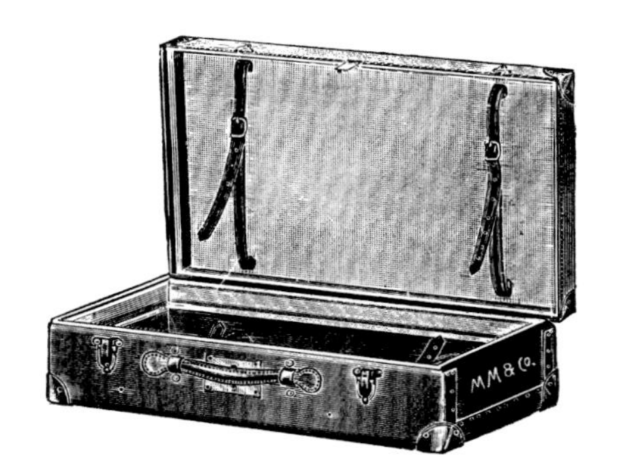

SIZES 22 in., 24 in.

No. 10 Dress Suit Case. Olive monkey grain and brown pebbled leather, lined with mauve imitation leather lining, straps in body, brass lock and catches.

No. 14 Dress Suit Case. Russet and brown grain leather, lined with imitation leather lining, straps top and body, brass lock and catches.

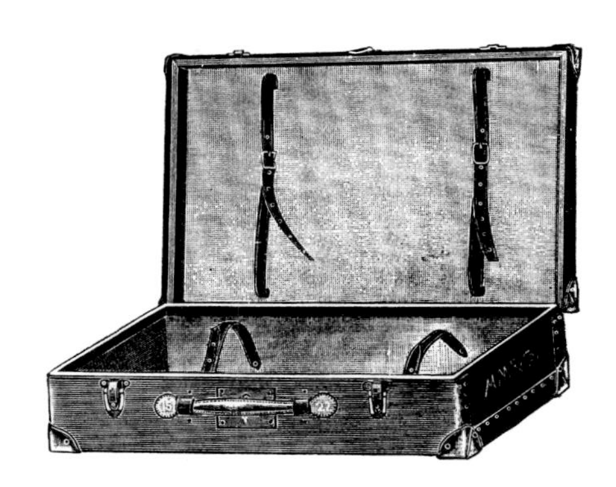

No. 15 DRESS SUIT CASE.

SIZES, 22 in., 24 in.

Russet grain leather, steel frame in body, Irish linen lined, straps in top and body, brass catches and lock.

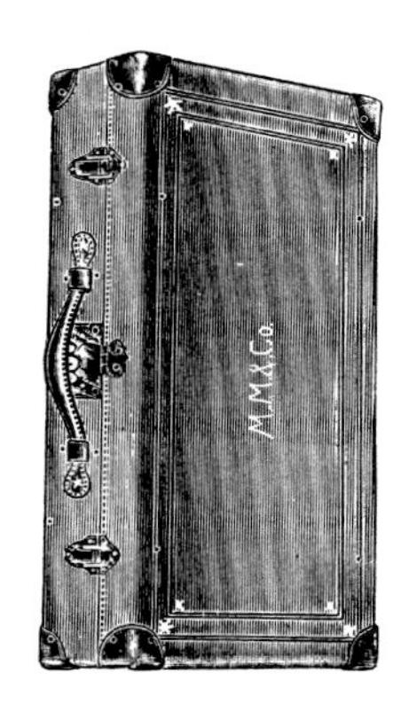

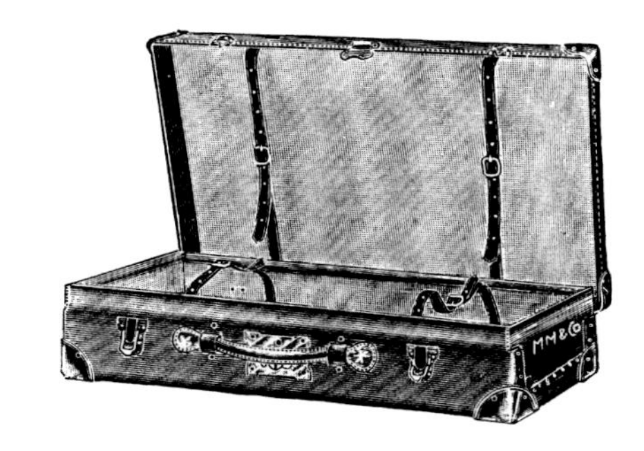

No. 16 DRESS SUIT CASE.

SIZES, 22 in., 24 in.

Fine cream or russet grain leather, steel frame, best catches and lock, Irish linen lined, straps top and body.

No. 1.

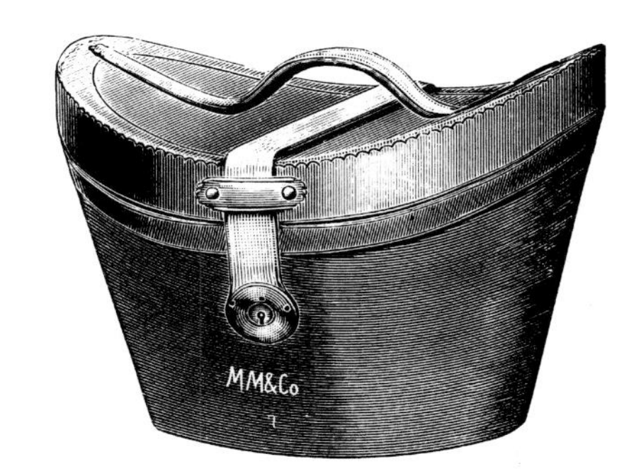

No. 2.

GENTS' SILK HAT BOXES.

No. 2. Grain leather covered, padded top and rest inside for hat, brass lock.

No. 1. Heavy grain leather, hand stitched bottom and top, silk lining padded in top, form for hat with padded rest, ornamental silk stitching, secure brass lock.

GROCERY CASE.

SIZES, **14 in., 16 in., 18 in., 20 in., 22 in.**

Made in russet, orange or brown pebble grain leather, nickel corners and shoes, hardwood cleats, leather covered on bottom, two doors. Bag lined throughout with Irish linen.

No. 3 DRESS SUIT CASE.

SIZES, 22 in., 24 in.

Duck covered, leather corners, cloth lined, inside straps, English handle nickel lock.

Choice Selection of High Grade Suit Cases and Gladstone Bags

Every Case Is Guaranteed. In Quality of Workmanship and Material They Are Unexcelled.

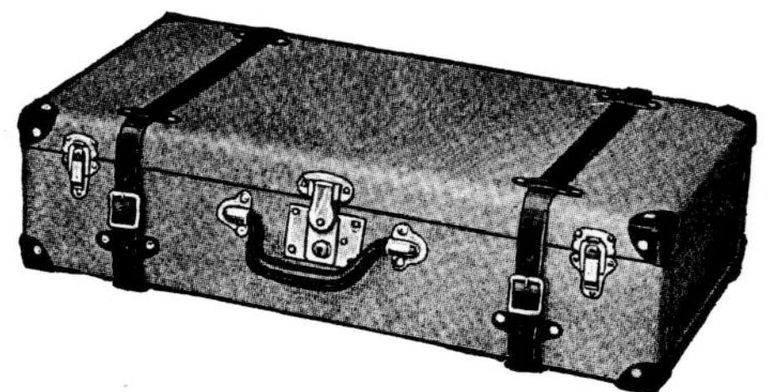

FIBRE MATTING SUIT CASE

Fibre matting, closely woven Brown japanned corners. Leather straps all around case.

	24-inch	26-inch
No. **1026**	**$5.10**	**$5.65**

Same without straps.

	24-inch	26-inch
No. **1016**	**$4.25**	**$4.65**

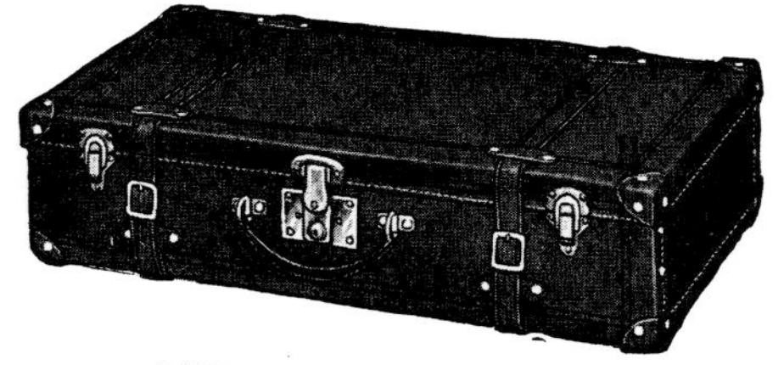

LOW PRICED SUIT CASE

"Lethr-Fibr." Good lock and handle. Straps all around. Brown japanned steel corners Neatly lined.

	24-inch	26-inch
No. **1096** Brown.............	**$4.65**	**$4.95**

Same without straps.

	24-inch	26-inch
No. **1076** Brown.............	**$3.50**	**$3.80**

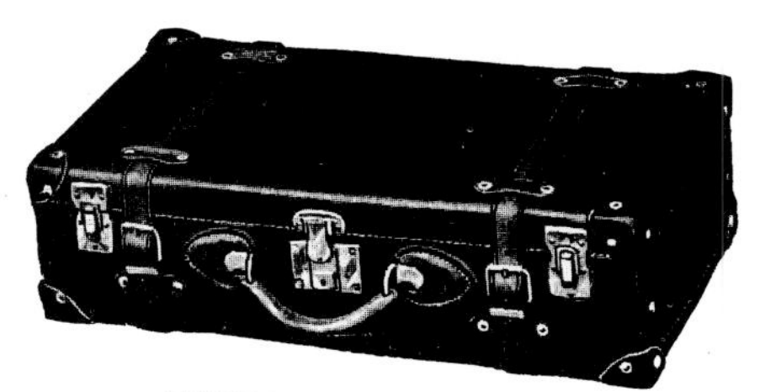

EXTRA DEEP SUIT CASE

"Lethr-Fibr." Extra deep case. Heavy leather corners. Straps all around. Ring handle. Cloth lined, and shirt pocket.

	24-inch	26-inch
No. **1126** Brown	**$9.95**	**$10.65**
No. **1146** Black	**9.95**	**10.65**

LADIES' SUIT CASES

Light in Weight, Perfectly Finished. Attractively Lined. Best Quality Throughout.

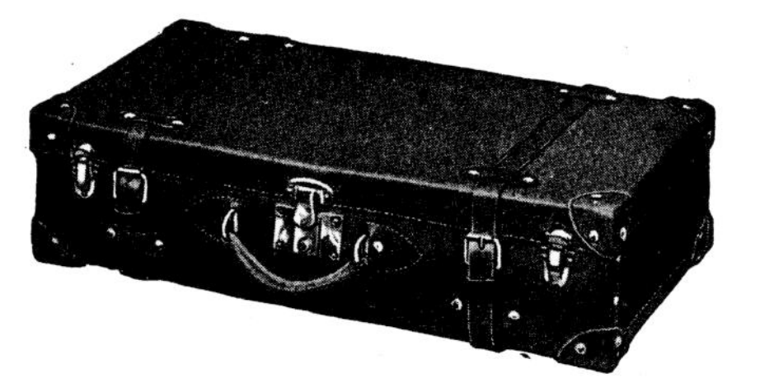

A GOOD SUBSTANTIAL SUIT CASE

Made of good, durable leather in crepe grain. Heavy leather corners. Good lock and catches. Heavy cowhide straps all around. Cretonne lined, with shirt fold.

		24-inch	26-inch
No. **1216**	Brown	**$18.40**	**$19.60**

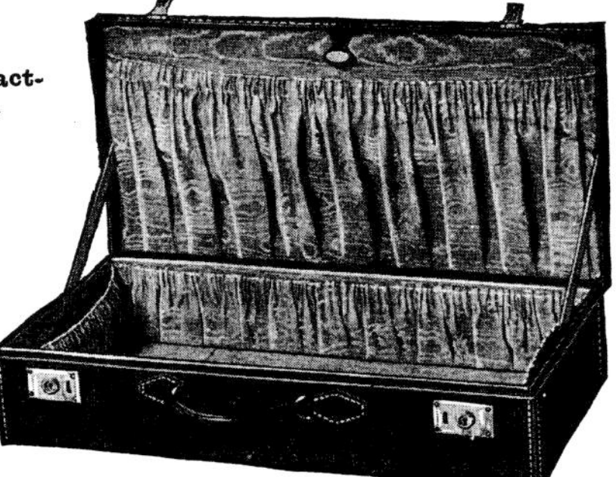

Light weight, attractive ladies' suit case, constructed of selected black cowhide in Windsor grain. Sewed French edges and sewed leather hinge, running the entire length of the case. Solid brass side lever lock. Attractive, silk moire lining. Soft shirt fold in lid and shirred pockets on ends.

	20-inch	22-inch	24-inch
No. **1596**	**$47.60**	**$50.40**	**$51.80**

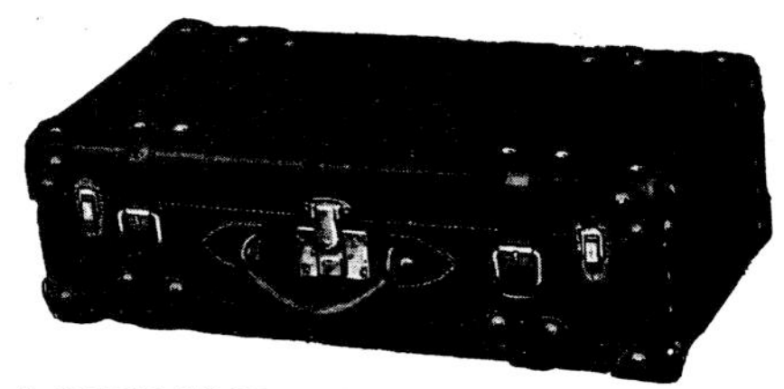

A VERY GOOD DURABLE LEATHER CASE

Made of good, durable leather. Extra deep. Leather corners. Bell rivets. Good lock and catches. Heavy cowhide straps all around. Linene lined, with shirt pocket.

		24-inch	26-inch
No. **971406**	Brown	**$18.45**	**$21.00**
No. **71456**	Black	**18.45**	**21.00**

$6.85 LADIES' LIGHT WEIGHT HAND MADE CASE.

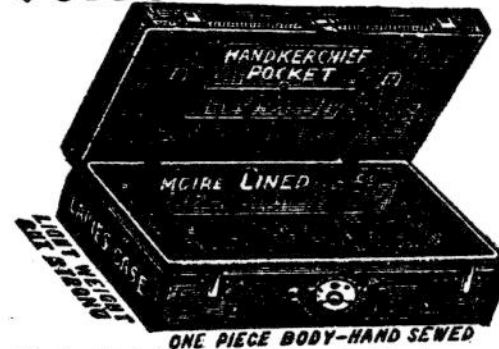

No. 33K1315 **Ladies' Light Weight Case of finest selected bridle leather in handsome russet shade.** One-piece body, hand made, beautifully hand creased and edges French finished. Satin finished solid brass lock and catches made especially for ladies' finest cases. Moire linen lined and **has** handkerchief pocket in cover and straps in bottom of case. Made light weight for ladies' use. **State size wanted.**

Length, 22 inches. **Our special price, only........$6.85**
Length, 24 inches. **Our special price, only........ 7.15**

$5.85 SOLE LEATHER SUIT CASE WITH STRAPS. PRICES REDUCED.

No. 33K1313 Heavy Selected Cowhide Suit Case, hand creased surface, sewed with waxed linen thread, finished with heavy bell riveted sole leather corners, round padded leather handle, brass lock and catches, brass rivets throughout, reinforced with two heavy straps and lined with Holland linen. Strapped shirt fold in top. Straps in bottom and at back to hold cover when open. Color, rich olive brown.

Length, 24 inches. **Reduced price, each, now only, $5.85**
Length, 26 inches. **Reduced price, each, now only, 6.15**

$4.85 COWHIDE SUIT CASE WITH SHIRT FOLD. PRICES GREATLY REDUCED.

No. 33K1316 This is the Greatest Value Ever Offered in a Genuine Bridle Leather Cowhide Suitcase. Selected stock strongly sewed with waxed linen thread and made over strong steel frame, linen lined and has linen shirt fold in cover closed with leather straps, also straps in bottom of case and stay straps at back to hold cover when open. Brass lock and catches, bell rivets, padded leather handle and sole leather corners. Artistically creased surface. Color a beautiful olive brown shade. We have materially reduced the price of this case this season, and we are able to supply an even better and handsomer case than before. **Please state size wanted.**

Length, 24 inches. **Reduced** price, now only.......**$4.85**
Length, 26 inches. **Reduced** price, now only....... **5.10**

$3.15 CHALLENGE OFFER GENUINE LEATHER DRESS SUIT CASE.

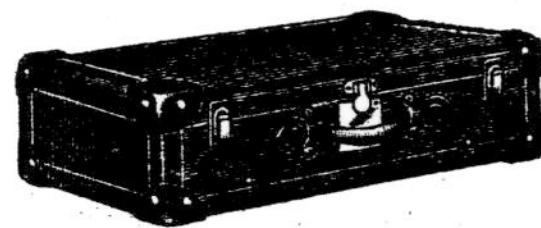

No. 33K1311 Genuine Leather Dress Suit Case of full weight stock, made over strong steel frame, brass riveted and sewed with waxed linen thread. Highly polished and handsomely creased surface. Has brass lock and catches, three hinges, round padded leather handle, solid cowhide riveted corners and is linen lined with full set of leather straps inside. A genuine leather case, which retails at **$4.50** to **$5.00**. Sole leather color only. We have improved the quality of the stock in this case and now challenge anyone to equal it at **$1.00** more than our low price. Length, 24 inches. Our price, now only.............................**$3.15**

$7.95 STRONGEST, MOST ELEGANT BRIDLE LEATHER CASE. EXTRA DEEP. PRICES REDUCED.

No. 33K1323 Strongest and handsomest Cowhide Bridle Leather Suit Case, highly finished and beautifully trimmed and hand creased, plaid linen lined. Heavy sole leather corners with six extra large brass bell rivets at each corner. Heaviest solid brass lock and catches, three brass hinges, and reinforced with two continuous cowhide straps, running through wide riveted leather loops. Padded solid leather handle. Linen shirt fold in cover, closed with two straps, two straps in bottom and stay strap at back, which holds cover of case when open. Color, the popular olive brown. The deepest and undoubtedly the strongest and one of the handsomest suit cases made. **State size.**

Length, 24 inches. Depth, 7½ inches. **Reduced** price, **$7.95**
Length, 26 inches. Depth, 7½ inches. **Reduced** price, **8.30**

$6.60 FOR FINE LEATHER LINED SUIT CASE. PRICES REDUCED.

From Photo.

No. 33K1320 Handsome Brown Cowhide Leather Suit Case of heaviest, highly polished bridle leather, made over strong steel frame, finished with heavy brass bell rivets in sole leather corners and on ends, sewed throughout with waxed linen thread. Handsomely creased surface. Solid brass lock, catches and hinges. Padded leather handle. Full leather lined body with leather shirt fold closed with two straps, also straps in bottom, and at back to hold cover. State size.
Length, 24 inches. Our reduced price, now only ...$6.60
Length, 26 inches. Our reduced price, now only ... 6.85

98c IMITATION LEATHER WATERPROOF SUIT CASE.

No. 33K1302 Brown Enamel Rubber Cloth Suit Case, solid cowhide leather riveted corners, reinforced round padded leather handle. Made over strong steel frame and riveted throughout. brassed lock and bolts, cloth lined and has inside straps; length, 24 inches; weight 6 pounds.
Our price, each, only 98c

$1.67 BEST IMITATION LEATHER SUIT CASE. PRICE REDUCED.

No. 33K1304 Best Waterproof Imitation Brown Grain Leather, Cloth Lined Suit Case, strongly riveted and sewed throughout, made on strong steel frame and has heavy riveted sole leather corners, round padded handle brassed lock and catches, and three strong hinges. Best and handsomest imitation leather case. Length, 24 inches. Reduced price, each, only..$1.67

22c LEATHER RIVETED CANVAS COVERED TELESCOPE.

No. 33K1344 Excellent Quality Canvas Covered Telescope. Full size. Corners leather reinforced. Padded and riveted leather handle and three cowhide straps on large sizes. State length wanted.

Length	Width	Weight	Height	Extd.	Price
14 in.	7 in.	1½ lbs.	6 in.	12 in.	22c
18 in.	9 in.	2½ lbs.	7 in.	14 in.	42c
20 in.	10 in.	3 lbs.	7½ in.	15 in.	52c
24 in.	12 in.	4 lbs.	8½ in.	17 in.	72c

$10.65 ONE-PIECE BODY, HAND MADE CASE.

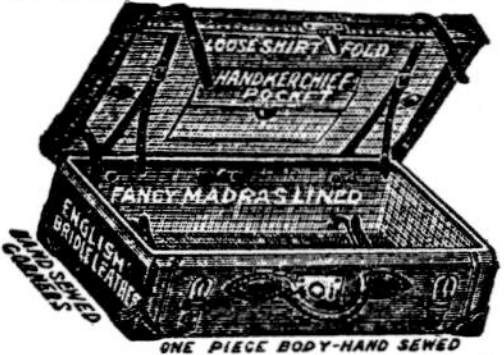

ONE PIECE BODY-HAND SEWED

No. 33K1324 The finest Suit Case made. Highest grade 5-ounce English bridle leather, popular russet color, one-piece body, hand sewed throughout, beautifully hand creased, finished with French edges, hand sewed corners, straps and handle and with solid brass lock of latest design. Instead of the usual catches near the ends of case, this case has hand sewed buckled straps; three strong brass hinges and four large bell rivets in bottom corners. Case is lined with popular plaid Madras and top contains removable shirt board held in place by two leather straps and finished with bellows pocket for handkerchiefs, etc. Straps in bottom and strap at end to hold cover when open. This is the most elegant and the highest grade cowhide suit case ever made and retails at $15.00 to $18.00. **State length wanted.**
Length, 24 inches. Our remarkable price, only....$10.65
Length, 26 inches. Our remarkable price, only.... 11.00

$9.35 FOR ENGLISH BELLOWS COWHIDE CASE. RETAILS AT $12.00 TO $14.00.

Reduced Prices.

5 OZ. COWHIDE BELLOWS CASE

HOLDS TWICE AS MUCH

No. 33K1322 English Bellows Case. Made of rich olive brown cowhide leather, selected stock, hand creased and sewed with waxed linen thread. Double bellows side, reinforced with two heavy cowhide straps running through broad riveted leather loops, finished with solid sole leather corners, secured with heavy bell rivets. Solid brass lock and catches, padded round leather handle. Lined with finest Holland linen. Strapped partition so that bellows side can be packed entirely separate from balance of case. Straps in bottom. **Be sure to state length wanted.** PRICES REDUCED AS FOLLOWS:
Length, 24 inches; weight, 10¼ pounds. **Reduced price** $ 9.35
Length, 26 inches; weight, 11¼ pounds. **Reduced price** 9.85
Length, 28 inches; weight, 12¼ pounds. **Reduced price** 10.35

$6.75 CONVENIENTLY FITTED COWHIDE LEATHER SUIT CASE. PRICE REDUCED.

SELECTED COWHIDE

No. 33K1317 Extra Heavy Selected Cowhide Suit Case, hand creased and sewed with waxed linen thread, riveted throughout and finished with best brass lock, catches and hinges, heavy bell riveted cowhide corners and padded leather handle, best linen lining. Has inside straps, including stay straps to hold cover when open and is fitted with hair brush, comb, soap box, tooth and nail brush in glass case and perfume bottle, each article firmly held in place by strong loops. A convenient and handsome suit case. Color, olive brown. Length, 24 inches. Reduced price, now only $6.75

$1.37 FOR BEST IMITATION ALLIGATOR SUIT CASE.

No. 33K1307 Finest Imitation Alligator Suit Case, highly finished surface, and can hardly be distinguished from genuine alligator. Reinforced round padded sole leather handle, riveted ends and brass plated lock and catches, strong steel frame, inside straps, cloth lined and waterproof. Length, 24 inches; weight, 7 pounds. Reduced price, only............$1.37

$2.15 MATTING SUIT CASE, STRAPS ALL AROUND. LIGHT WEIGHT BUT SERVICEABLE.

No. 33K1305 Popular Olive Matting Suit Case, leather and leatherette trimmed and with straps all around. Very closely woven and strong Japanese matting braided or woven in such a manner that it is practically waterproof and has the additional advantage of being so light in weight that a child can easily carry it. Made over strong steel frame and finished with riveted cowhide corners, edges and ends neatly bound with leatherette. Strong brassed lock, round padded leather handle and two strong leather straps running all around. One of the most practical cases ever made. Length, 24 inches. This case retails at $3.00. Our price, each, only....$2.15

49c FULL LEATHER BOUND CANVAS TELESCOPE. PRICES REDUCED.

No. 33K1346 Strongest Waterproof Canvas Telescope. Leather bound ends. Strongly sewed and riveted corners and ends, and broad sewed and riveted handle. Finished with three solid grain leather straps on large sizes. State length wanted.

Length	Width	Weight	Height	Extd.	Reduced Prices
16 in.	8¼ in.	2 lbs.	6½ in.	12 in.	$0.49
20 in.	10¼ in.	3 lbs.	8 in.	14 in.	.69
24 in.	12¼ in.	4 lbs.	9½ in.	17 in.	.89
26 in.	13½ in.	4½ lbs.	10½ in.	18½ in.	1.09

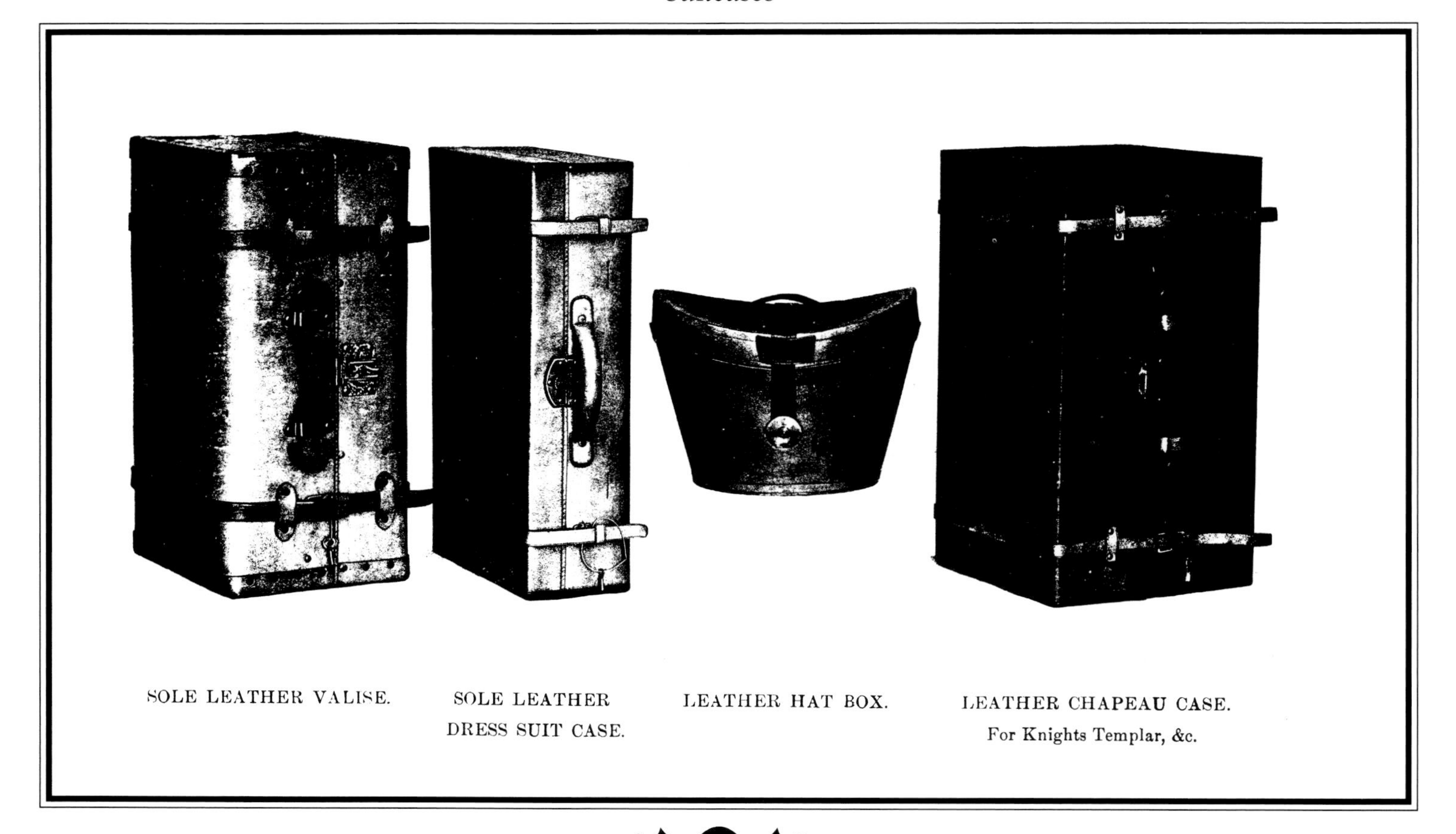

SOLE LEATHER VALISE.

SOLE LEATHER DRESS SUIT CASE.

LEATHER HAT BOX.

LEATHER CHAPEAU CASE. For Knights Templar, &c.

TELESCOPE CASES

No. 1 TELESCOPE.

No. 5 SUIT TELESCOPE.

SIZES, 16 in., 18 in., 20 in., 22 in., 24 in., 26 in.

Leather binding rivetted, leather corners, good straps and handle cloth lined, handle rivetted to bar on under side.

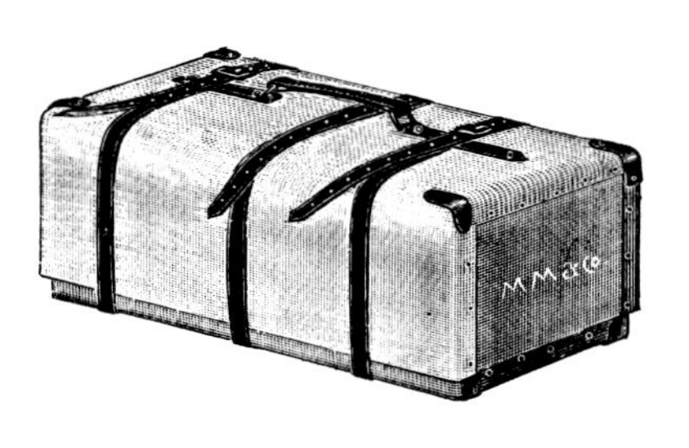

No. 9 TELESCOPE.

No. 4 SUIT TELESCOPE.

No. 9.	**SIZES,**	12 in.,	14 in.,	16 in.,	18 in.,	20 in.,	22 in.,	24 in.,	26 in.
No. 4.	**SIZES,**	16 in.,	18 in.,	20 in.,	22 in.,	24 in.,	26 in.		

Cloth lined, good straps; handle riveted to bar on under side, leather tips.

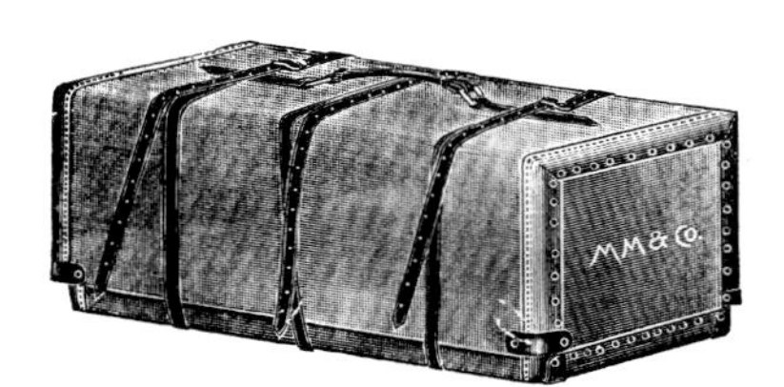

No. 6 TELESCOPE.

No. 8 SUIT TELESCOPE.

No. 6.	**SIZES,**	14 in.,	16 in.,	18 in.,	20 in.,	22 in.,	24 in.,	26 in.
No. 8.	**SIZES,**	16 in.,	18 in.,	20 in.,	22 in.,	24 in.,	26 in.	

Leather binding on edges stitched and rivetted, cloth lining, good straps and handles rivetted to bar on under side.

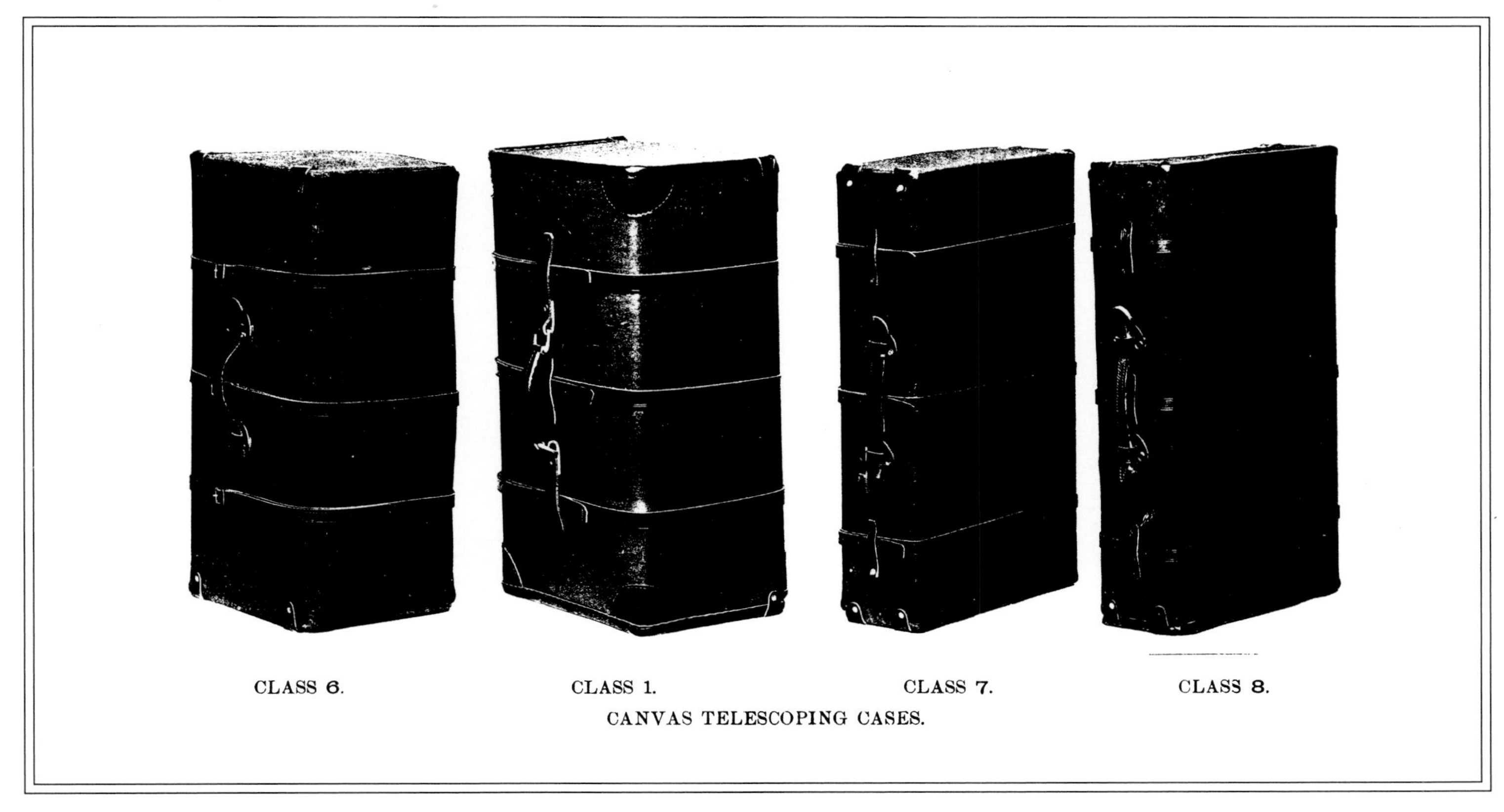

CLASS 6. CLASS 1. CLASS 7. CLASS 8.

CANVAS TELESCOPING CASES.

Brief Cases

INSURANCE CASE

A very convenient case that is reinforced throughout and will prove highly serviceable. Made of Cowhide Leather in Brown or Black colors, smooth finish. It has 3 pockets of which one pocket has a 3-inch expansion, and 2 pockets each 1½-inch expansion. Stock size 12x6 inches. 3 pockets.

No. 2V251A Black Color, each..........................$7.90
No. 2V251C Brown Color, each.......................... 7.90

STANDARD SIZES ARE
LEGAL SIZE, 15x10 INCHES.
MUSIC SIZE, 16x11 INCHES.
BE SURE TO SPECIFY COLOR WANTED WHEN ORDERING.

BRIEF CASES

Made to order in five sizes and choice of 1, 2, 3 or 4 pockets. A size for every requirement and convenience.

Made of 4½ to 5 oz. Cowhide Leather in Brown or Black colors, smooth finish. Expert workmanship and fine finish.

Be sure to specify Color wanted.

No.	Size		Pockets	Each
2V103A	12x 8	inches	One	$ 6.15
2V103B	12x 8	inches	Two	7.50
2V103C	12x 8	inches	Three	8.80
2V103D	12x 8	inches	Four	10.15
2V103E	14x 9½	inches	One	6.95
2V103F	14x 9½	inches	Two	8.30
2V103G	14x 9½	inches	Three	9.60
2V103H	14x 9½	inches	Four	10.95
2V103J	15x10	inches	One	7.75
2V103K	15x10	inches	Two	9.10
2V103L	15x10	inches	Three	10.40
2V103M	15x10	inches	Four	11.75
2V103N	16x10½	inches	One	8.40
2V103P	16x10½	inches	Two	9.75
2V103R	16x10½	inches	Three	11.10
2V103S	16x10½	inches	Four	12.40
2V103T	18x12	inches	One	11.20
2V103U	18x12	inches	Two	13.20
2V103V	18x12	inches	Three	15.20
2V103W	18x12	inches	Four	17.20

PORTFOLIOS

Choice of 15-inch or 16-inch sizes. Made of 4½ oz. Cowhide leather in Brown or Black Colors. Smooth finish. Remarkably well made, reinforced, giving most satisfactory service.

Be sure to state Color wanted.

The construction of 15-inch size is as follows:
- 2 Pockets, 15x10x1½ inches.
- 2 Pockets, 8x6x1 inches.
- 2 Pockets, 7½x5 inches.
- 2 Pencil loops.

The construction of 16-inch size is as follows:
- 2 Pockets, 16x10½x1½ inches.
- 2 Pockets, 8x6x1 inches.
- 2 Pockets, 8x5 inches.
- 2 Pencil Loops.

PRICES

No. 2V37A 15-inch size, each..........................$18.00
No. 2V37B 16-inch size, each.......................... 20.70

CATALOG CASES

Made to order in five sizes—a size for every requirement.

Made of 5 oz. Cowhide Leather in Brown or Black, smooth finish. Bottom is stiffened and reinforced. Case is divided into 2 large pockets. On Leather Partition are 2 small pockets, suitable for order book, envelopes, pencils, etc.

Be sure to specify Color wanted.

No.	Size	Each
2V200A	12x 8 x2½ inches	$ 8.40
2V200B	15x10 x3½ inches	14.00
2V200C	16x10½x4 inches	16.80
2V200D	17x11 x4½ inches	21.10
2V200E	18x12 x5 inches	25.35

BRIEF CASES

Made to order with choice of sizes and 3, 4 or 5 pockets.

A size for every requirement.

Made of 4½ to 5 oz. Cowhide Leather in Brown and Black smooth finish. Equipped with Adjustable Handle.

Be sure to specify Color wanted.

No.	Size	Pockets	Each
2V229A	14x 9½ inches	Four	$13.90
2V229B	15x10 inches	Three	13.10
2V229C	15x10 inches	Four	14.40
2V229D	15x10 inches	Five	15.75
2V229E	16x11 inches	Three	14.40
2V229F	16x11 inches	Four	15.75
2V229G	16x11 inches	Five	17.10

CATALOG CASES

Made to order with choice of four Stock sizes—A size for every requirement.

Made of 5 oz. Cowhide Leather in Brown or Black colors, smooth finish. Gussets with weather-proof flaps. Bottom is stiffened and reinforced. Case is divided into 3 large pockets with Leather Partitions. On first partition are two flat pockets for order or memo. book, envelopes, pencils, etc.

Be sure to specify Color wanted.

No.	Size	Each
2V225A	15x10 x4 inches	$18.00
2V225B	16x10½x4½ inches	20.70
2V225C	17x11 x5 inches	25.10
2V225D	18x12 x5½ inches	29.35

SATCHELS

No. 4 CLUB.

No. 6 CLUB.

SIZES, 10 in., 11 in., 12 in., 13 in., 14 in., 15 in., 16 in.

No. 4 Club. Duck, with leather corners, Jap. frame, cloth lined, nickel trimmings.
No. 6 Club. Pebbled split leather, Jap. frame, cloth lined, nickel trimmings.
No. 8 Club. Imitation alligator leather, Jap. frame, cloth lined, nickel trimmings.

SIZES, 10 in., 11 in., 12 in., 13 in., 14 in., 15 in., 16 in,

No. B Club. Pebbled leather, imitation leather lined, Jap. frame, English handle, nickel trimmings, spring catches on frame.

No. C Club. Pebbled grain leather, imitation leather lining, Jap. frame, English handle, nickel trimmings.

No. W Club. Pebbled leather, leather lined, Jap. frame, spring catches, English handle, nickel trimmings.

No. D Club. Pebbled grain leather, leather lined, Jap. frame, English handle, nickel trimmings.

No. P CLUB.

SIZES, 10 in., 11 in., 12 in., 13 in., 14 in., 15 in., 16 in.

No. P Club. Pebble grain leather, covered frame, leather lined, nickel trimmings.

No. O Club. Same style and finish as P. Club, with stay hinge attachment.

No. E CLUB.

SIZES, **12 in., 13 in., 14 in., 15 in., 16 in., 17 in., 18 in.**

Made in fancy olive grain leather, covered frame, covered inlays, stay hinge, leather lined, fine nickel trimmings.

No. S CLUB.

SIZES, 12 in., 13 in., 14 in., 15 in., 16 in., 17 in., 18 in.

No. S Fancy colors, grain leather, covered frame and inlays, fine leather lining, two inside pockets, improved patent sliding lock, patent automatic catches and stay hinges, hand stitched handles, best nickel trimmings.

No. S Genuine seal leather, handsome silk lining, otherwise made and trimmed as above.

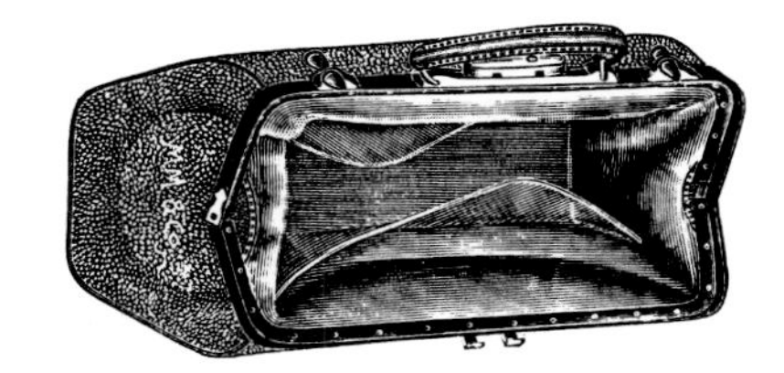

OBSTETRICAL.

SIZES, 14 in., 15 in., 16 in., 17 in., 18 in.

Black and brown pebbled grain leather, chamois lined, covered frame and inlays, nickel trimmings, spring catches on frame.

No. M CLUB.

SIZES, 12 in., 13 in., 14 in., 15 in., 16 in., 17 in., 18 in.

Alligator leather, leather lined, covered frame, spring catches; nickel trimmings, stay hinge.

EXPRESS BAG.

BANK BAG.

SIZES, **14 in.,** **15 in.,** **16 in.**

Express Bag—Covered frame, grain leather bottom, linen lined.

Bank Bag—Has safety steel frame, leather covered; heavy duck sides stitched to frame; strap leather bottom, linen lined.

No. 2 MONEY SAC.

SIZES, 7 in., 8 in., 9 in.

Pebbled grain leather, duck lined, Jap frame, nickel trimmings.

No. 64 CLUB.

SIZES, 12 in., 13 in., 14 in., 15 in., 16 in., 17 in., 18 in.

Fancy colors, grain leather, fine leather lining, nickel trimmings, pocket inside, stay hinge.

SIZES, 12 in., 13 in., 14 in., 15 in., 16 in., 17 in., 18 in.

No. 76 Harvard. Olive grain leather; imitation leather lined, Jap. frame, spring catches, brass trimmings.

No. 72 Harvard. Olive grain leather; leather lined. spring catches, Jap. frame, brass trimmings.

No. 74 Harvard. Olive grain leather; leather lined, covered frame, spring catches, stay hinge, nickel trimmings.

No. 33 HARVARD.

SIZES, 14 in., 15 in., 16 in., 17 in., 18 in.

No. 33. Brown or black pebbled grain, plain leather lining, three inside pockets.

No. 33. Fancy colors grain leather, pig lining, three inside pockets.

No. 33. Alligator, pig lining, three inside pockets. All styles have English lock, spring catches, Columbus handles. Brass trimmings, stay hinge.

No. 35 HARVARD. **No. 60 HARVARD.**

SIZES, 14 in., 15 in., 16 in., 17 in., 18 in.

No. 35 Harvard. Fancy colors grain leather, leather lining, covered frame and inlays, special English handle, English lock, snap catches, nickel trimmings, stay hinge.

No. 60 Harvard. Fancy colors grain leather. Has no welt or corner seam. Patent applied for. Covered frame, pig lined, three inside pockets, Columbus handle, stay hinge.

No. 60 Harvard. Alligator. Made and finished same as the grain bag.

No. 27 OXFORD.

SIZES, 12 in. to 18 in.

No. 30 HARVARD.

SIZES, 14 in. to 18 in.

No. 27 Oxford. Fancy colors grain leather, sewed in frame; fine leather lining, three pockets inside, stay hinge, Columbus handle, brass trimmings.

No. 27 Oxford. Alligator, same make and finish as above.

No. 30 Harvard. Fancy colors, grain leather, English sewed in frame, pig leather lined, three inside pockets, self-acting stay hinge, Columbus handle, either spring clasps or sliding catches.

No. 30 Harvard. Alligator, same make and finish as in grain leather.

No. 42 HARVARD.

No. 40 HARVARD.

SIZES, 14 in., 15 in., 16 in., 17 in., 18 in.

No. 42 Harvard. Fancy colors grain leather, English sewed in frame, fine leather lining, soft sides, Columbus handle, fine nickel trimmings, snap hinge.

No. 40 Harvard. Fancy colors grain leather, English sewed in frame, leather lined, inside pocket, stay or self acting hinge.

No. 40 Alligator Harvard. Made and finished as the grain bag.

No. 45 HARVARD.

SIZES, 14 in., 15 in., 16 in., 17 in., 18 in.

Fancy Colors Grain Leather and Alligator.

English frame, sewed in; hand stitched edges, no welt; pig leather lined, three inside pockets, Columbus handle, fine nickel trimmings.

No. 65 HARVARD.

SIZES, 14 in., 15 in., 16 in., 17 in., 18 in.

In Fancy Colors Grain Leather and Alligator.

Patent applied for. Has no welt or corner seams. English sewed in frame, pig leather lined, three inside pockets, stay hinge, Columbus handle, fine nickel trimmings.

No. 1 X CABIN.

SIZES, 12 in., 13 in., 14 in., 15 in., 16 in., 17 in., 18 in., 20 in.

No. 1 X. Brown pebbled grain leather, covered frame, leather lined, end pockets, English handle, nickel trimmings.

No. 1 X. Fancy colors grain leather, Columbus handle; otherwise same as above.

No. 1 X. Alligator, pig leather lined; otherwise made and finished same as 1 X fancy.

No. 4 X CABIN.

SIZES, 12 in., 13 in., 14 in., 15 in., 16 in., 17 in., 18 in.

Brown pebbled grain leather, Jap. frame, imitation leather lining, end pockets, nickel trimmings.

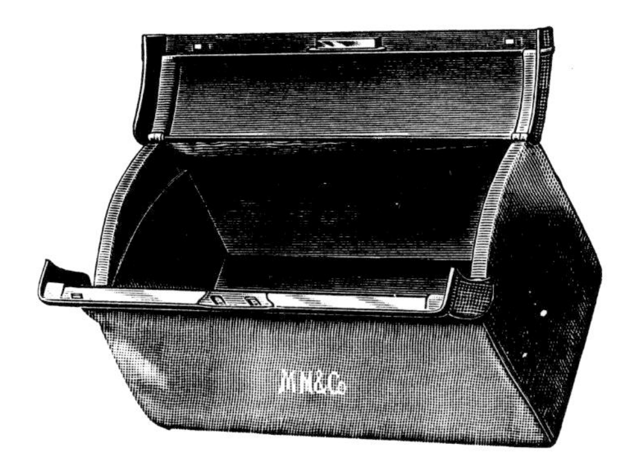

No. 7 CABIN.

SIZES, 12 in., 13 in., 14 in., 15 in., 16 in., 17 in., 18 in.

Made in Brown Pebble Grain, Fancy Colors Grain and Alligator Leathers.

English sewed in frame, either sliding or spring clasp top catches, Columbus handle, pig leather lined, end pockets.

No. 3 CABIN.

SIZES, 10 in., 11 in., 12 in., 13 in., 14 in., 15 in. 16 in., 17 in., 18 in.

No. 3 Cabin. Brown pebbled grain leather, covered sewed-in frame, leather lined, two inside pockets, English handle.

No. 3 Cabin. Fancy colors grain leather, Columbus handle; otherwise same make and finish as above.

No. 3 Cabin. Alligator, pig lined; otherwise finished as No. 3 fancy.

No. 3 Seal. Made in genuine black seal; otherwise finished as No. 3 Alligator.

No. 8 CABIN.

SIZES, 12 in., 13 in., 14 in., 15 in., 16 in., 17 in., 18 in.

Fancy colors grain leather. Patent applied for. Has no welt or corner seams. English sewed-in frame, pig leather lined, end pockets, Columbus handle.

Philip Schild & Co., who have succeeded Kaufman, Schild & Co., report a prosperous outlook. They make a specialty of alligator traveling bags and have just gotten out the "Prince of Wales" an illustration of which appears herewith. This bag has taken Europe by storm

and is evidently a seller in this country. Made in all leathers with imported frames, they are exceedingly "chic" in appearance, of great durability. They also have a new ladies cabin bag made in grain and alligator leathers and of first class stock throughout which they claim to be offering to the trade at the lowest price a reliable article was ever before sold.

HAT CASE (Made in two sizes)
(Description to right)

Description of Trunks and Hat Cases as follows:

Dark blue vulcanized fibre covering with walnut fibre binding. Five-ply construction. Panels studded with saddle nails. Brass plated steel trimmings. Spring lock and draw bolts. Fancy cloth faced. Full cloth lined. Four drawers for linen, hats, etc. Top drawer divided into 3 compartments. Hat crown in bottom drawer. Shoe pockets. Ratchet bar follower. Size: Height, 40 inches; width, 23½ inches; depth, 22 inches

No. W739 Full size Fibre Covered Wardrobe, each........$71.50

Dark green vulcanized fibre covered. Black vulcanized fibre binding. Five ply construction. Brass plated steel trimmings. Spring lock. Draw bolts and double dowel plate. Fancy cloth faced. Drawers full cloth lined. Double trolley. Full set of garment hangers. Ratchet bar follower. Height 40 inches; width 22 inches; depth, 22 inches.

No. W703 Round Edge Full Size Wardrobe, each.........$53.35

Black vulcanized fibre covering with black vulcanized binding. Five ply construction. Brass plated steel trimmings. Draw bolts and spring lock. Fancy design cloth faced. Four cloth lined drawers. Patented clothes retainer. Size: Height, 40 inches; width, 13¾ inches; depth, 22 inches.

No. W709 Fibre Covered Steamer Wardrobe, each........$50.70

De Luxe Traveling—(Nothing Carried)

A new mode of traveling has been introduced in this country by the convenience and comfort the taxi service renders. All that is now necessary is to call your taxi; have chauffeur take you, Wardrobe trunk and Hat case along. Upon arrival at depot **"George will do the rest."**

Black enameled duck Hat case. Bound with leather on all wearing edges. Nickel hardware. Fancy cretonne lining with shirred pocket in lid. Four shirred pockets in the body of case. Cretonne faced tray two hat crowns.

No. W81218 Hat Case, size 18x18 inches, each...........$31.35
No. W81220 Hat Case, size 20x20 inches, each........... 32.70

BOSTON BAGS MADE IN THREE QUALITIES

Very neat in appearance. Lining is of fancy blue cretonne which is glued to the sides. Construction of black, fine grain, patent finished Dupont Fabrikoid. Over-lap sewed frame. Stitched end seams. Double handle. Cowhide crepe grain strap with buckle fastener. 15-inch only.

No. W7 Each..$4.00

Medium Grade

Flexible inner stock cowhide leather. Double stitched end seams. Serviceable cloth lining. One pocket. Double handle. Heavy straps and brass roller buckle fastener. Sewed in steel frame.

Sizes	14-Inch	15-Inch	16-Inch
No. W16 Brown, each	$5.35	$6.00	$6.70
No. W18 Black, each	5.35	6.00	6.70

Hand Boarded Cowhide Boston Bag (Highest Grade)

Made of flexible grained cowhide leather. Construction same as above.

Sizes	14-Inch	15-Inch	16-Inch
No. W26 Brown, each	$10.70	$11.35	$12.00
No. W28 Black, each	10.70	11.35	12.00

SUIT CASES AND TRAVELING BAGS

Only the finest grades and the toughest and best wearing qualities of selected grain leather. Made over strong steel frames and reinforced underneath with heavy binders' board.

$5.85 BLACK WALRUS GRAIN COWHIDE OXFORD BAG.

No. 33K1238 Hand Sewed Black Walrus Grain Oxford Bag, leather lined. Only the best selected cowhide bag leather of full 6-ounce weight is used in this bag, but the embossed grain on the leather is such a perfect imitation of genuine walrus that an expert can hardly distinguish the difference. Lined with serviceable buff color leather with pocket inside, and has sewed and riveted corners. English round padded leather handle and strong handsome brassed lock and catches. Large and roomy. **State size wanted.**

Length, 16 inches. Cut price......**$5.85**
Length, 18 inches. Cut price...... **6.45**

59c BROWN SPLIT LEATHER BAG.

No. 33K1210 Heavy Split Cowhide Leather Club Style Bag. Looks like genuine brown grain leather. Has cloth lining, round padded leather handle, brassed lock and catches and large bell rivets in bottom. Big value at our low price. **Be sure to state length wanted.**

Length, 10 inches. **Special cut price, $0.59**
Length, 12 inches. **Special cut price, .79**
Length, 14 inches. **Special cut price, .95**
Length, 16 inches. **Special cut price, 1.12**

$2.15 GENUINE GRAIN LEATHER BAG.

No. 33K1230 Genuine Russet Grain Leather Bag, in the popular Oxford shape, cloth lined. A medium priced leather bag that will give excellent service. Welted seams sewed with waxed linen thread, brassed lock and catches and has five large bell rivets on bottom. The best medium priced genuine leather bag ever shown. **State size.**

Length, 14 inches. **Special price....$2.15**
Length, 16 inches. **Special price.... 2.55**
Length, 18 inches. **Special price.... 2.95**

$3.45 GRAIN LEATHER OXFORD BAG. PRICES REDUCED.

No. 33K1232 Fine Selected Heavy Grain Leather Oxford Bag, with good leather lining and inside pocket. A sightly bag and will give best of service. Welted seams sewed with waxed linen thread. Brass lock and catches, keratol covered steel frame, English round padded leather handle and heavy bell rivets on bottom. **State size wanted.**

PRICES REDUCED AS FOLLOWS:

Length, 14 inches. **Reduced price, now only..........................$3.45**
Length, 16 inches. **Reduced price, now only..........................$3.90**
Length, 18 inches. **Reduced price, now only..........................$4.35**

$9.95 BEST GENUINE BLACK WALRUS TRAVELING BAG.

No. 33K1244 **Extra Large Shape, First Quality Genuine Walrus Oxford Bag.** Genuine black walrus leather distinguished by its deep heavy grain, is absolutely waterproof and the best and handsomest bag leather known. Will wear a lifetime. Sewed with strongest waxed linen thread. Hand made over imported English frame. Lined with best buff color leather and has capacious inside pocket. Hand sewed walrus corners, solid brass locks and catches of the strongest and latest improved pattern. Lock protected by large walrus flap with slot for card or name plate. This handsome black walrus traveling bag is unequaled for service and appearance by any bag retailed at $18.00 to $20.00. **State size wanted.**

Length, 16 inches. **Our special price, only......$ 9.95**
Length, 18 inches. **Our special price, only...... 10.95**
Length, 20 inches. **Our special price, only...... 11.95**

$8.95 FOR GENUINE BRIDLE LEATHER TRAVELING BAG. PRICES REDUCED.

No. 33K1251 Heavy Highly Finished Genuine Bridle Leather Oxford Bag. Made of 6-ounce selected cowhide, welted seams sewed with waxed linen thread. English frame which opens up full width. Lined with fine buff color selected leather and finished with large inside pocket. Round padded English leather handle. Satin finished, solid brass catches, and lock covered with leather flap with slot for name plate or card. Large bell rivets on bottom. Large, roomy shape and will wear forever. Retails at $15.00 to $18.00. **Please state size.**

Length, 16 inches. **Our reduced price, now only..$ 8.95**
Length, 18 inches. **Our reduced price, now only.. 9.95**
Length, 20 inches. **Our reduced price, now only.. 10.95**

No. 34 BROWN RUBBER.
Imitation Alligator.

No. 80 BROWN LEATHER.
No. 81 BLACK LEATHER.

No. 103 LEATHER. Imitation Alligator.
No. 104 LEATHER. Imitation Alligator.
Leather Lined.

No. 180 BROWN GRAIN LEATHER.
Leather Lined.
No. 182 BROWN GRAIN LEATHER, Cloth Lined.

No. 304 BROWN GRAIN LEATHER.
Leather Lined.

No. 306 DARK GRAIN LEATHER.
Imitation Alligator. Leather Lined.

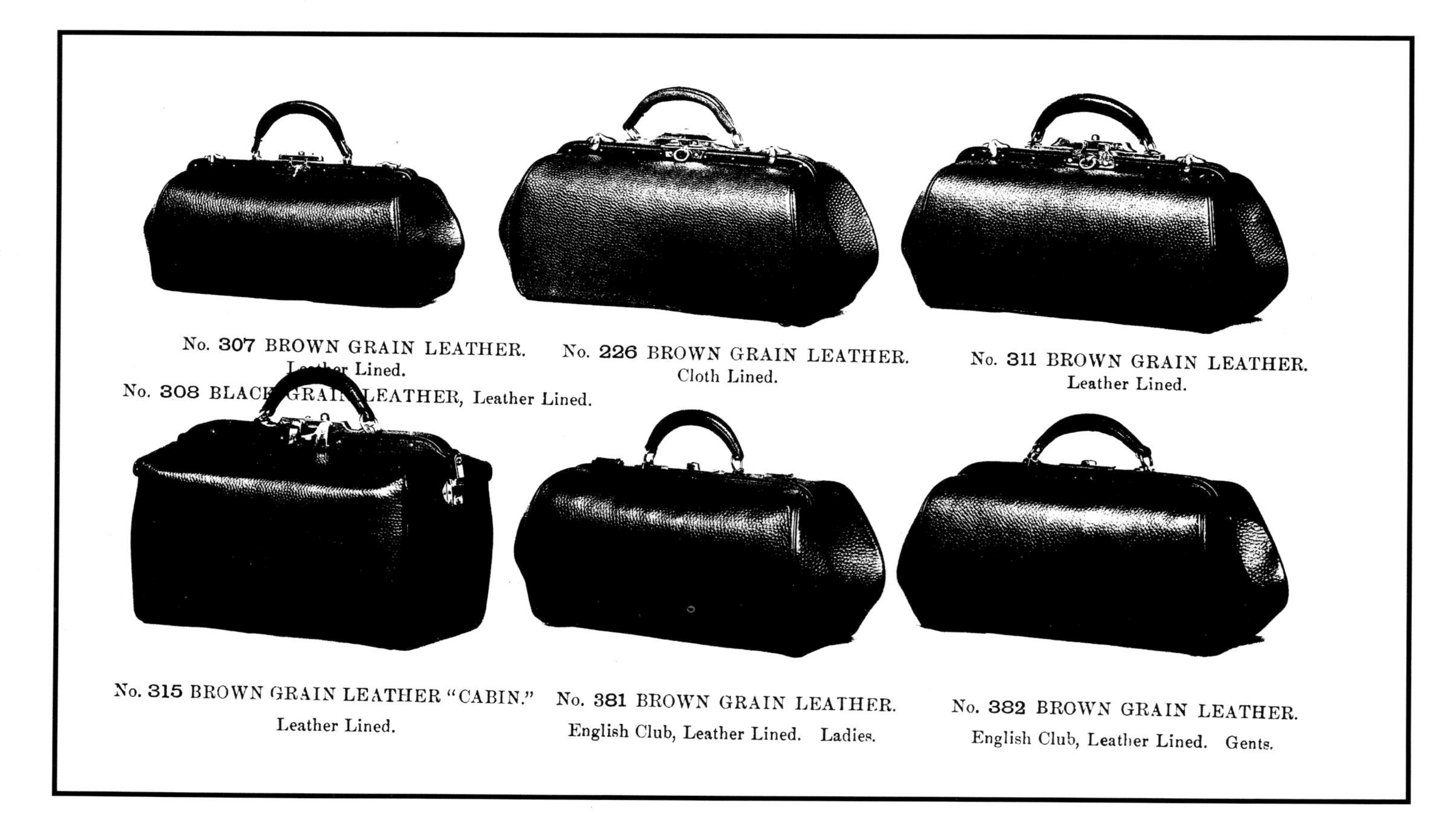

No. 307 BROWN GRAIN LEATHER.
Leather Lined.
No. 308 BLACK GRAIN LEATHER, Leather Lined.

No. 226 BROWN GRAIN LEATHER.
Cloth Lined.

No. 311 BROWN GRAIN LEATHER.
Leather Lined.

No. 315 BROWN GRAIN LEATHER "CABIN."
Leather Lined.

No. 381 BROWN GRAIN LEATHER.
English Club, Leather Lined. Ladies.

No. 382 BROWN GRAIN LEATHER.
English Club, Leather Lined. Gents.

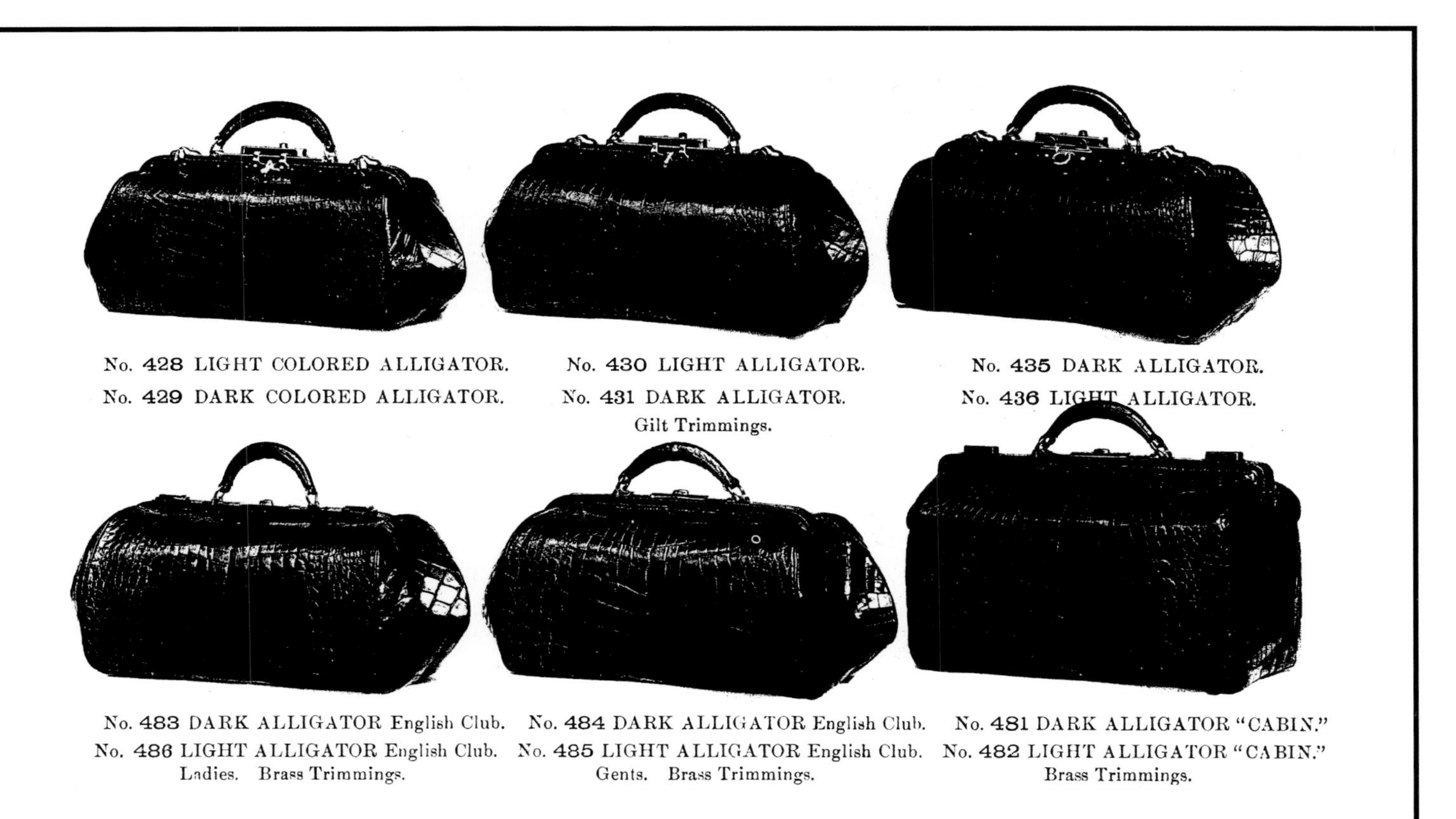

No. 428 LIGHT COLORED ALLIGATOR.
No. 429 DARK COLORED ALLIGATOR.

No. 430 LIGHT ALLIGATOR.
No. 431 DARK ALLIGATOR.
Gilt Trimmings.

No. 435 DARK ALLIGATOR.
No. 436 LIGHT ALLIGATOR.

No. 483 DARK ALLIGATOR English Club.
No. 486 LIGHT ALLIGATOR English Club.
Ladies. Brass Trimmings.

No. 484 DARK ALLIGATOR English Club.
No. 485 LIGHT ALLIGATOR English Club.
Gents. Brass Trimmings.

No. 481 DARK ALLIGATOR "CABIN."
No. 482 LIGHT ALLIGATOR "CABIN."
Brass Trimmings.

NOVELTIES

THE TRUNK AND LEATHER NOVELTIES REVIEW.

Vol. 5. No. 5 May (12), 1894.

FINE POCKET-BOOKS,
CARD and LETTER CASES

Sterling Silver Mountings a Specialty.

DANIEL M. READ,

156 Greene Street, cor. Houston, NEW YORK.

WM. A. HAINES,

POCKET-BOOKS, CARD AND LETTER CASES, CUFF AND COLLAR BOXES, &c.

•••••••

133, 135 & 137 NORTH SEVENTH ST.,

PHILADELPHIA.

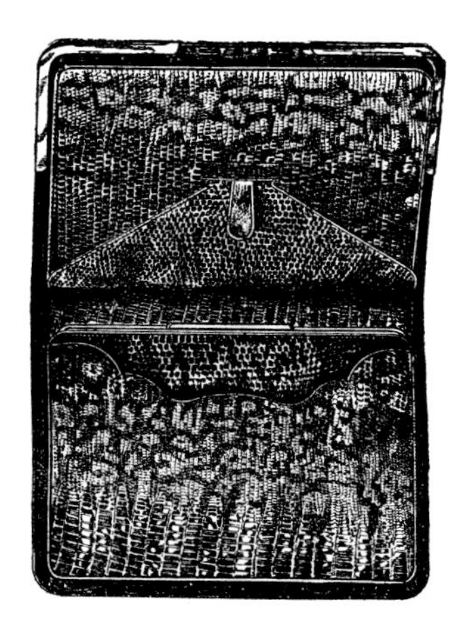

C. F. RUMPP & SONS,

Manufacturers and Importers of

FINE LEATHER GOODS

Pocket Books, Purses, Card Cases, Cigar Cases, Cuff and Collar Boxes, etc., etc.,

STERLING SILVER MOUNTED GOODS.

Fifth and Cherry Streets, PHILADELPHIA

NEW YORK SALESROOM, - 621 BROADWAY, (The Cable Building)

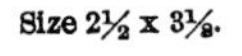

Size 2½ x 3⅛.

THE SAFETY SPECIE PURSE AND POCKET-BOOK.

The Most Convenient Purse and Pocket-Book Made.

Specie cannot lose out. Contents seen at a glance. Soft and flexible in the pocket. Made in two sizes, small size for Specie, large size for Specie, Bills, etc. Dealers in Leather Novelties will find these goods popular sellers. Samples sent to the trade. Write for circular and price-list.

Arms Pocket Book Co. Manufacturers of FANCY LEATHER GOODS.

HARTFORD, CONN.

Size 2½ x 3⅝.

Patented January 30, 1892.

Unique Pat. Combination Bill Fold & Coin Purse.

Separate places for coin, bills and car tickets, independent of each other; Flexibility, Lightness; No metal parts to get out of order, or wear the pocket. Send for Catalogue.

No.												
No. 11	holds	$ 4.00	in silver,	10	Notes	and Car Ticket,	$ 3 00	Doz.	$ 6.00	Doz.	$ 8.00	Doz.
" 16	"	6.00	"	20	"	"	6 co	"	10 50	"	13 50	"
" 13x	"	9.00	"	12	"	"	6 75	"	13 00	"	15 00	"
" 13½x	"	10.00	"	20	"	"	7.50	"	15 00	"	19 50	"

James Topham, Sole Mfr., 1231 Pa. Ave., Washington, D.C.

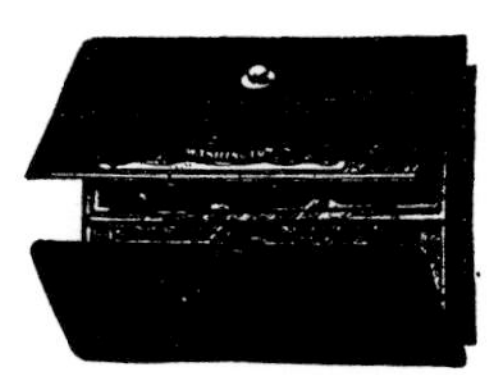

Patented January 30, 1892.

H. M. ROSENBLATT & CO.,
MANUFACTURERS,
237, 239, 241, 243, 245, 247 and 249 Monroe Street, CHICAGO, ILL.
SEND FOR NEW ILLUSTRATED PRICE LISTS.

EGGEMAN'S
Collar and Cuff Portfolio.

Points in which it excels all others:

It takes up less room in the bag.
It is more convenient to pack.
Does not injure the finish of a collar or cuff.

The ends being flexible, it adapts itself to the inner shape of a bag, be it large or small.

14 inch, English Gladstone.

Patented Jan. 30, 1894.

13 inch Club.

The Eggeman Collar and Cuff Portfolio is made in Sheep, Morocco, Grain, Russian Calf, Seal and Alligator. Seal and Alligator is leather lined sold at from $9 to $48 per dozen. ***A sample order solicited.*** You will be pleased with them.

G. S. EGGEMAN, 315 SUMMIT ST., TOLEDO, O.

A FINE ASSORTMENT OF BANKERS' CASES.

F. RUMPP & SONS, Philadelphia, have this season added several new designs to their line of railroad pass cases, one of which is herewith illustrated. These goods are made in a variety of leathers, comprising grain, seal, alligator and lizard, all of uniform excellence in workmanship and finish. Samples of the Key-

In the assortment of bankers' cases are also a number of new combinations, one of which, the extension back and adjustable buttonlock, is shown in cut. In this case the pockets can be filled to their utmost capacity and yet the cover may be opened or closed without any difficulty. The annoyance of bulging and consequent strain and wear is avoided, as the proper shape of the pockets is always retained.

These cases are made of grain cowhide with leather gussets and finished in the best possible manner to secure strength. Illustrated price lists will be mailed on application.

Bankers Case from C. F. Rumpp & Sons' Line.

TRUNK & LEATHER NOVELTIES

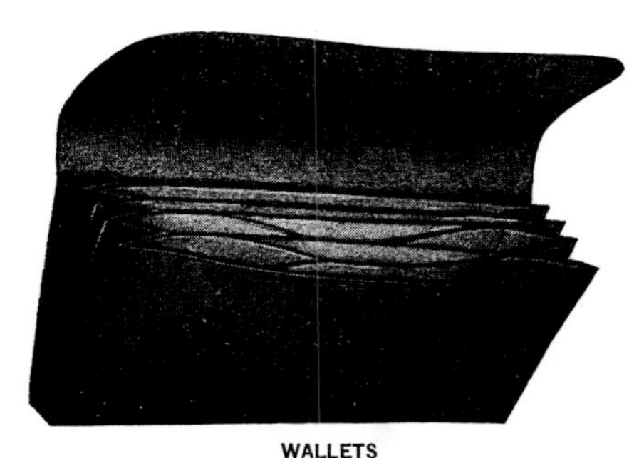

WALLETS

A size for every requirement.

Made of Brown or Black finish leather. Well made and elegantly finished. Priced very low.

Be sure to specify Color wanted.

No.	Size	Pockets	Each
2V15A	10-Inch	One	$1.50
2V15B	11-Inch	One	2.15
2V16A	10-Inch	Two	2.15
2V16B	11-Inch	Two	2.80
2V17A	10-Inch	Three	3.35
2V17B	11-Inch	Three	4.30
2V18A	10-Inch	Five	4.55
2V18B	11-Inch	Five	5.60

MUSIC ROLLS (Sheet Music Size)

No. 2V153 Made of Heavy Imitation Leather, Black finish, each $1.50

No. 2V154 Made of Heavy Split Cowhide Leather, Black Seal Grain finish, each 2.30

No. 2V155 Made of Cowhide Leather in Brown, Black Smooth and Black Seal Grain finish, each 3.50

No. 2V156 Made of Hand Boarded Cowhide Leather in Brown and Black finish, each 2.80

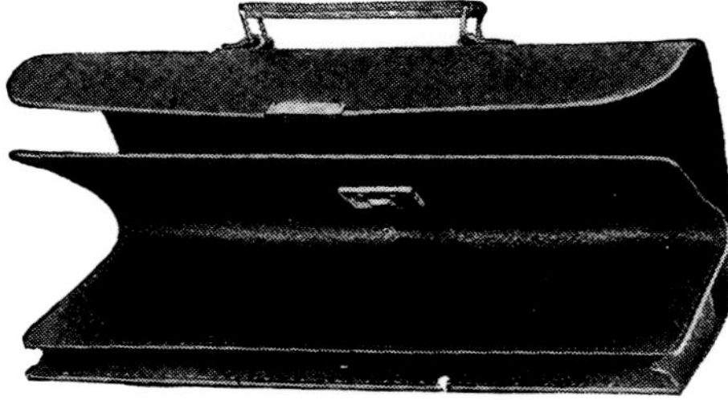

MUSIC CASES (Sheet Music Size)

No. 2V115 Made of Cowhide Leather in Brown, Black Smooth finish. Can be carried full size or folded, each $8.55

No. 2V116 Same as above but made of Split Cowhide Leather, each 6.15

No. 2V117 Same as above but made of Imitation Leather, each 4.00

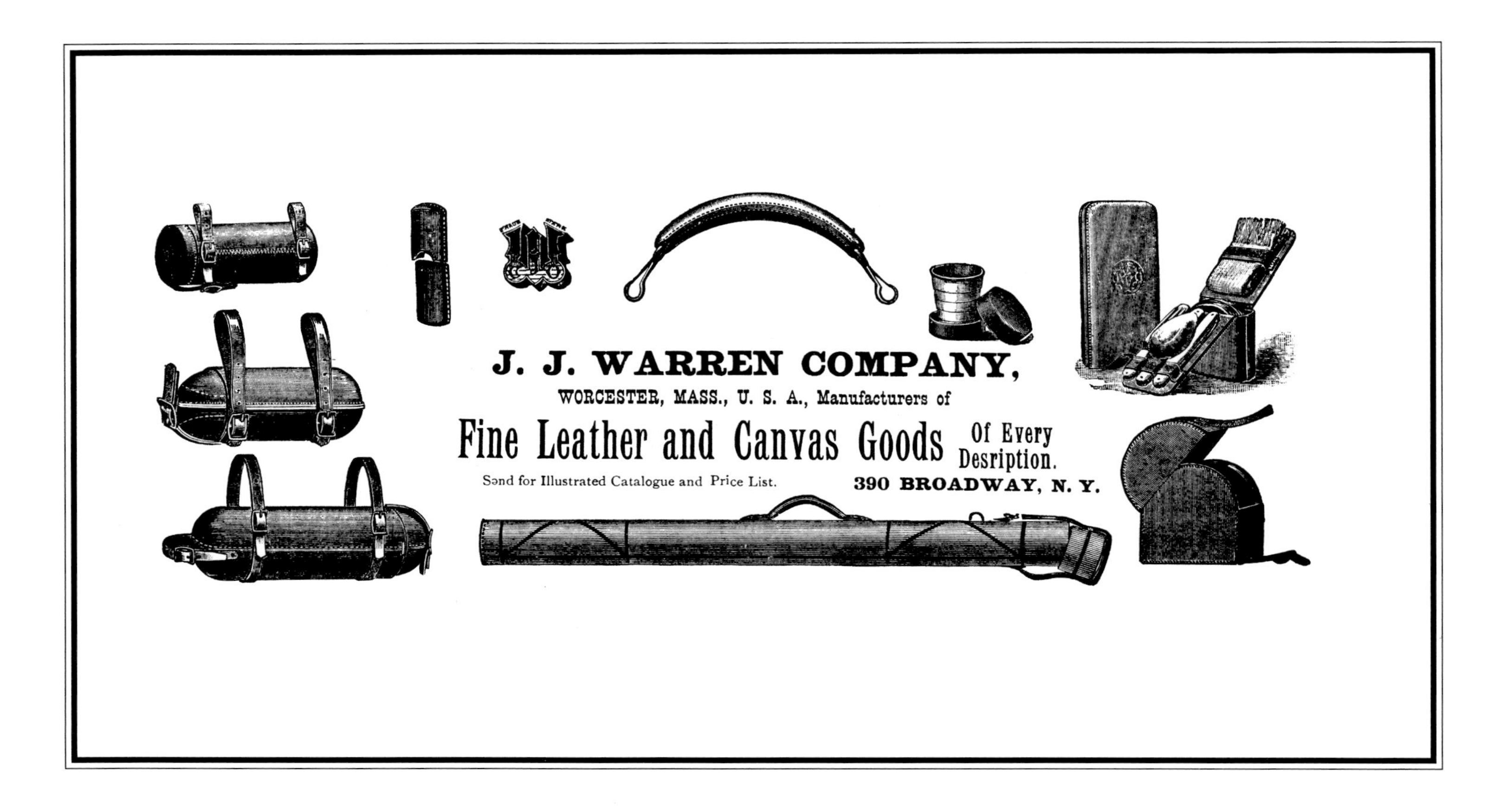
J. J. WARREN COMPANY,
WORCESTER, MASS., U. S. A., Manufacturers of
Fine Leather and Canvas Goods Of Every Desription.
Send for Illustrated Catalogue and Price List.
390 BROADWAY, N. Y.

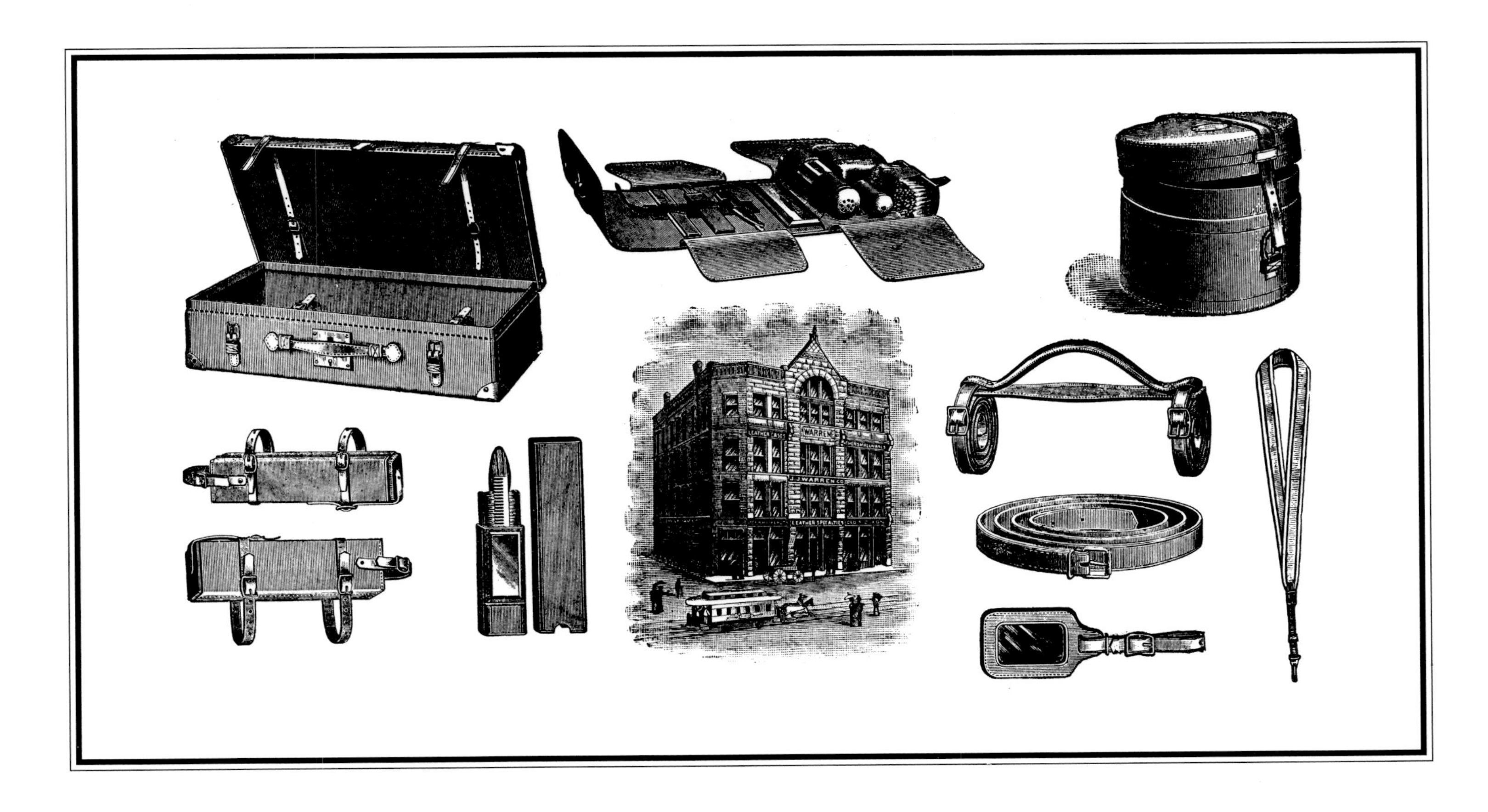
LEATHER CASES
WARREN
J. J. WARREN CO
LEATHER SPECIALTIES

Patent Applied For.

Market prices vary considerably from one part of the country to another. Determining value of a particular piece depends on size, condition, materials from which it was made, and overall quality of design. Trunks, for example, are also judged on completeness, such as whether they still contain their original compartments and trays. If the original inserts are missing, the value is lowered accordingly.

This guide is only intended to be a general reference. The prices listed for trunks refer to those that are judged complete.

Trunks:

~Metal covered	Barrel-top	$375 & up
	Flat -top	$225 & up
~Wood	Saratoga	$450 & up
	Flat-top	$300 & up
~Fiber	Flat-top	$125 & up
~Canvas-covered	Saratoga	$225 & up
	Flat-top	$175 & up
	Steamer	$150 & up
~Leather covered	Saratoga	$400 & up
	Flat-top	$300 & up
	Steamer	$275 & up
	Imitation leather	$150 & up
~Kartavert	Fiber	$125 & up
~Wardrobes	Wardrobe trunk	$300 & up
	Wall dresser trunk	$375 & up

Traveling Bags:

~Gladstones	Leather	$100 & up
	Imitation leather	$50 & up
	Rubber	$50 & up
	Canvas covered	$45 & up
~Suitcases	Leather	$85 & up
	Imitation leather	$45 & up
	Canvas covered	$40 & up
	Fiber	$35 & up
~Telescope cases	Leather	$100 & up
	Canvas	$50 & up
~Brief cases	Leather	$95 & up

Satchels:

Leather	$85 & up
Imitation leather	$45 & up
Cloth covered	$35 & up